# *Arduinos Without Tears*

Book 1 in the *Saturday Afternoon Low-Cost Electronics Projects* Series

Eric Bogatin

Addie Rose Press

*Second Edition*

September 2019

Copyright 2019 by Eric Bogatin

All rights reserved

ISBN: Color Version, Paperback, 978-1-7325670-4-7

ISBN: B&W Version, Paperback: 978-1-7325670-5-4

A Publication of Addie Rose Press

## *Preface: Buy this ebook*

More than 1 million Arduinos have been sold in the last 10 years. They are enabling kids from 8 years old to 80 years old to enter the exciting world of physical computing; the marriage of science, electronics and computers. Arduinos can make all of your projects "smart."

If you want to get started, or just want to dip your toe to see what all the buzz is about, this eBook is for you.

If you've tried to figure out Arduinos from a web site, another eBook, or an online course, but struggled in getting started, this eBook is for you.

If you already use an Arduino or are an engineer or scientist wantabee and want to develop good habits and skills you will use for the rest of your career, this eBook and my other eBooks are for you.

All the experiments, projects and examples in this eBook can be completed *with just an Arduino*. There are *no other components or parts you need*. I show you where you can buy an Arduino for less than $4. This is an incredibly small investment with a terrific return on investment.

You will not find an easier to follow, more in-depth introduction to Arduinos, electronics and science, for as low an investment as this eBook.

*Welcome to the wonderful world of Arduinos.*

Check out Eric Bogatin's other video eBooks in the *Saturday Afternoon Low-Cost Electronics Projects* series:

*Science Experiments with Arduinos Using a Multi-Function Board*

*More Science Experiments with Arduinos Using an LED Kit*

*Science Experiments with the SparkFun Redboard Turbo*

*Three Must Have Instruments for Every Lab*

*Electronics You Should Have Learned in High School*

You can find more details about these and other resources on his author's page on Amazon.com and his author's website, and the HackingPhysics web page.

Better yet, subscribe to the Hacking Physics Journal.

*Drop him a note sometime and let him know what you think!*

## Table of Contents

Chapter 1.   Why the Arduino is For You ................................... 11
   1.1   Why this video eBook is for you ................................... 12
   1.2   Purchase ALL the parts you will need for this eBook .... 17

Chapter 2.   Talking to Your Arduino ......................................... 19
   2.1   What you need to know and what you will learn in this chapter ................................................................................... 19
   2.2   Your first step ................................................................. 20
   2.3   The Quick Start ............................................................... 20
   2.4   Step 1: Purchase an Arduino for less than $4 ................. 22
   2.5   Step 2: Download the Arduino IDE ................................ 31
   2.6   Step 3: Launch the IDE and a blank sketch is opened .... 34
   2.7   Introducing the Syntax of a programming language ...... 37
   2.8   Summary of important vocabulary ................................. 41
   2.9   Using the help menu to get started ................................. 42
   2.10   Step 4: Connecting your Arduino for the first time .... 43
   2.11   Step 5: Select the board and com port ........................ 46
   2.12   Troubleshooting: no COM port or it's grayed out and nothing to select ....................................................................... 49
   2.13   Step 6. Upload a blank sketch .................................... 51
   2.14   Troubleshooting- what if it didn't work? Enter Forensic Analysis 54
   2.15   Step 7: Congratulations! ............................................. 56
   2.16   Summary of the new skills you have learned in this chapter   56

Chapter 3.   Getting to Blink ...................................................... 58

3.1    What you need to know and what you will learn in this experiment .................................................................................. 58

3.2    The Quick Splash ............................................................. 59

3.3    Setting preferences and a good habit .............................. 62

3.4    A quick tour of the Arduino board ................................... 69

3.5    Command: pinMode(), the first command to bend the Arduino to our will. ....................................................................... 73

3.6    New feature: the built-in Reference, where you can find all the commands ....................................................................... 75

3.7    New Command: digitalWrite(), Turning on or off output voltages ................................................................................... 78

3.8    New command: delay() ................................................... 80

3.9    New Code: use comments ............................................... 82

3.10   My Sketch: Blink ........................................................... 84

3.11   Summary of the commands introduced so far ............... 84

Chapter 4.    It's alive .................................................................. 87

4.1    What you should know before you start this chapter ...... 87

4.2    Giving your Arduino a heartbeat .................................... 87

4.3    My Sketch: It's Alive ....................................................... 89

4.4    Printing "Hello World" ................................................... 90

4.5    New Code: Serial.begin() ............................................... 91

4.6    New Code: Serial.print() ................................................ 94

4.7    New Feature: The serial monitor .................................... 96

4.8    How fast can you print to the serial monitor? .............. 101

4.9    Summary of the commands introduced so far ............. 104

Chapter 5.    Modulating the On-Board LED and Persistence of Vision     107

5.1   What you need to know and what you will learn in this experiment ................................................................... 107

5.2   The basic pulse train pattern .......................................... 108

5.3   What is the flicker rate? ................................................. 110

5.4   Changing the apparent brightness ................................. 113

5.5   New Features: Introducing variables and variable types 114

5.6   New Features: Variable names ..................................... 121

5.7   On-time, off-time, period and duty cycle ..................... 125

5.8   My sketch to modulate the LED with a period and duty cycle   128

5.9   Summary of the commands introduced so far .............. 128

Chapter 6.   Plotting Patterns in Numbers ................................... 132

6.1   What you need to know and what you will learn in this experiment ................................................................... 132

6.2   Printing columns of data ............................................... 132

6.3   Generating a series of numbers .................................... 134

6.4   My sketch: incrementing an index and printing index, index$^2$ 135

6.5   An excel trick to place all the data into separate columns 136

6.6   New Code: the if…else command ............................... 138

6.7   My Sketch to create a ramp up and down or a triangle pattern of numbers ..................................................... 140

6.8   New Feature: Plotting the data in the IDE with the Serial Plotter 143

6.9   New Feature: Auto format is your friend ..................... 147

6.10　Advanced: Controlling the number of decimal places in the Serial.print() command ...................................................... 150

6.11　Summary of the commands introduced so far ........... 152

Chapter 7.　How Random are Random Numbers? ................... 156

7.1　What you need to know and what you will learn in this experiment ...................................................................... 156

7.2　New command: the for loop ........................................ 156

7.3　Random numbers ......................................................... 159

7.4　Random walking .......................................................... 160

7.5　Estimating deviation likelihood from 0 in a random walk 162

7.6　An electronic coin flipper ............................................. 163

7.7　Calculating a random walk .......................................... 166

7.8　My Sketch: Plotting a random walk with limits ........... 166

7.9　A better metric for the worst-case likely deviation ....... 169

7.10　What if we keep increasing the number of coin flips? 170

7.11　My sketch: random walking with multiples of 500 coin flips　172

7.12　Relative distance traveled ......................................... 173

7.13　Can you win at Roulette using this simple method? . 177

7.14　My sketch to test out the sure bet approach to winning at roulette ........................................................................ 178

7.15　Summary of the commands introduced so far ........... 181

Chapter 8.　Measure Static Electric Fields .............................. 185

8.1　What you will need, commands you should know, commands you will learn ............................................... 185

| | | |
|---|---|---|
| 8.2 | Static electric fields or 60 Hz pickup | 186 |
| 8.3 | Turning electric fields into voltages with an antenna | 187 |
| 8.4 | New hardware feature: ADC | 187 |
| 8.5 | Taking your first ADC measurements | 189 |
| 8.6 | Storing ADC values in a variable in units of ADU | 192 |
| 8.7 | Averaging over n power line cycles (PLC) | 195 |
| 8.8 | My sketch: display measurements averaged over n power line cycles | 197 |
| 8.9 | Add an antenna to increase the sensitivity to stray electric fields | 201 |
| 8.10 | Measure static charges | 203 |
| 8.11 | How the ADC measures electric field | 206 |
| 8.12 | A simple static charge experiment with a cat | 210 |
| 8.13 | Static charge experiments | 212 |
| 8.14 | Electrostatic damage (ESD) | 215 |
| 8.15 | An advanced note | 217 |
| 8.16 | Summary of the commands introduced so far | 218 |
| Chapter 9. | Other examples: exploring other sketches | 222 |
| 9.1 | What you need to know and what you will learn in this experiment | 222 |
| 9.2 | Built-in examples | 223 |
| 9.3 | Six popular web sites with many sketches | 225 |
| Chapter 10. | Your Next Steps | 227 |
| Chapter 11. | An Introduction to This Book Series. | 229 |
| 11.1 | The name of this series of books says it all. | 230 |
| 11.2 | Welcome to the world of physical computing | 232 |

| 11.3 | All the books in this series | 233 |
| 11.4 | About Eric Bogatin | 234 |
| 11.5 | How to stay in touch | 235 |

# Chapter 1. Why the Arduino is For You

The Arduino is my workhorse lab assistant. It is the very first device I grab when I want to start any dynamic project that is going to interact with the world, any measurement project, any sensor project, or any control project.

If you are not sure if the Arduino is for you, do a Google search for "Arduino Projects" and you will see more than 2 million hits.

If this is too overwhelming, there are three web sites with thousands of cool projects for beginners to experts. If you want a glimpse of where you can go with Arduinos, check out the projects on these three web sites:

Sparkfun

Adafruit

Arduino.cc

In addition to the thrill of applying science and technology principles to the real physical world, I love working with Arduinos because they are a way of expressing my *creativity*. They are like any form of *art* or even *performance*, a medium to express my passion, my thoughts, my style and show off my talent.

> *Building projects with Arduinos or any sort of electronics projects is a chance to exercise your creativity. Every time you create something that has more than one path to get there, the result*

*reflects your decisions. Your personality comes through in how you implement your project.*

## 1.1 Why this video eBook is for you

Welcome to my video eBook, <u>Arduinos Without Tears</u>. It is really a video course packaged as an ebook. I provide four important features in this video eBook:

- *It's a standard book you can just read.*
- *It's available in eBook and PDF format so you can read it on any platform, even in a browser.*
- *All the code examples embedded in this video eBook can be copied out and pasted directly into a blank sketch and they will just work. You can also download zip files with code that is longer than a handful of lines.*
- *Each section has a video associated with it which you can view through any browser. These are unlisted videos so you can only access them through the links in this video eBook.*

This is like a video course. You don't really have to read the text, you can just browse to the video links and view the videos. If you want some more of the background, go ahead and read the text. Go at your own pace.

The **key feature** of this video eBook is that it lowers the barriers to entry to take Arduinos for a test spin. This video eBook is the:

- ✓ *absolutely simplest course*
- ✓ *for the absolutely lowest cost*

- ✓ with the absolutely lowest investment of time
- ✓ with no soldering
- ✓ no wiring
- ✓ no other components needed.

I've been teaching Arduino workshops to folks from 8 years old to 80 years old, some of whom never touched an Arduino before, and some who tried to get started but stopped in frustration. In Figure 1.1 is a picture of one of my workshops at Tinkermill, our hackerspace in Longmont, CO.

*Figure 1.1 An example of one of my Arduino workshops at Tinkermill in Longmont, CO.*

Many folks who come to my workshops have previously tried to get started with Arduinos but failed. They come to my workshop to give it one last chance. I keep hearing the same five reasons why they got frustrated and gave up:

1. It costs too much to get started.
2. There are so many different courses, which one do I choose?
3. It's way too hard with all those wires.
4. Every sketch I open looks so complicated.
5. I spent an hour and couldn't even get my Arduino talking to my computer.

I hear over and over again that when they've encountered a problem, they can never find the answer and the frustration level skyrockets with every forum or blog they read that doesn't address their specific question. It's not that there is no information on-line, there is *too much* and none of it is *understandable*.

I designed my video eBook, <u>Arduinos Without Tears</u>, specifically for the folks who have tried to get started with Arduinos but encountered one of these sorts of barriers, got frustrated and gave up.

> *The purpose of this video eBook is to eliminate these barriers. If you want to get started with the exciting world of Arduinos, these won't be in your way anymore.*

And, if you have already tried the Arduino and feel like this is the gadget for you, I'll show you a few tricks and best design practices, or habits, you won't find in many other guides.

In this introductory video eBook, <u>Arduinos without Tears</u>, your investment in money and time is less than $4 to buy an Arduino and a few Saturday afternoons of playing around. If you get excited with the projects and experiments, you could spend as long as 6-9 months exploring all of the options.

There is no wiring or soldering or additional components. All of the examples and all of the experiments in this video eBook use just the Arduino board, like the one shown in Figure 1.2.

*Figure 1.2 An example of the Arduino Uno board we will use for all the examples in this book, with its power light on.*

Granted, limiting ourselves to just the Arduino board by itself, there isn't a lot we can do. But you will get enough of a taste to know if physical computing - combining the physical world, the world of electronics hardware and coding - is an activity for you.

The essence of the Arduino, and physical computing in general, is to *control aspects of the physical world*. Generally, most projects with Arduinos, a special class of *microcontrollers*, involves taking some information from the outside world, converting it into a voltage that can be read by the Arduino and then, based on the code, or *sketch* we write, converting this input voltage to an output voltage to control something else. This general process is illustrated in Figure 1.3.

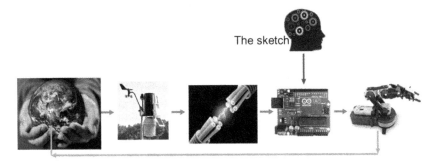

*Figure 1.3. The general process of using an Arduino to take information from the physical world, convert it, using a sketch, into an output to then control something else in the physical world.*

In this video eBook, we will see how to engineer and design the instructions to the Arduino to make the on-board LED jump through hoops. Then, we will use the Arduino to jump through some programming hoops and help us explore number-patterns, display them on a plotter or save them into text files. Finally, we'll use the on-board voltage sensor to measure the static electric fields surrounding every one of us, all the time.

One purpose of this course is to get you to the point where you can evaluate the world of Arduinos and judge for yourself that this is one of the coolest new worlds in which you will want to play.

*Another purpose of this video eBook is to jump start your skills to the next level, so you are armed with good habits and basic techniques you can use in all your future projects.*

For less than $4 and a few Saturday afternoons, you will have a realistic idea of how exciting the world of Arduinos can be and why there are more than 1 million other enthusiasts.

If you love the feeling of *empowerment* you have at your fingertips, a *powerful assistant* you can control, and a new outlet for your *creativity*, you are ready for the next step.

In the follow-on video eBooks in this series, we use add-on sensors and actuators which we manipulate with the Arduino to interact with the physical world. I selected the specific add-on components as low cost and those which require very little wiring and no soldering, yet give you immediate access to the exciting world of science and engineering.

> *And, remember, in this video eBook, you don't need to know any electronics, there is no soldering, no wiring, no extra components. We just use the Arduino board itself.*

*Welcome aboard.*

## 1.2 Purchase ALL the parts you will need for this eBook

All of the experiments in this book can be completed with just the simple Arduino board, USB cable and your computer: a PC or Mac. You will need NOTHING else. Here is a quick link to purchase an Arduino if you do not already have one:

|  | Arduino Uno with USB cable.<br><br>Price is about $3.00 to $3.75, shipping included.<br><br>Purchase from this link, for example.<br><br>If it is no longer available, you can do a search on www.aliexpress.com for Arduino Uno, and select any vendor. |
|---|---|

*Watch this video and I will walk you through where you can purchase a very low-cost Arduino board.*

# Chapter 2. Talking to Your Arduino

This is the chapter to start with if you have never worked with an Arduino. We start with downloading installing and running the IDE on your computer and plugging the Arduino into the USB port. By the end of this chapter your computer should be talking to your Arduino. It should take about 5 minutes.

If you are already there, you can skip this chapter.

## 2.1 What you need to know and what you will learn in this chapter

To begin this adventure, all you need to know is some basic PC or Mac skills like surfing the web, downloading a program, and installing it. You do not need to have any previous experience with programming.

The Arduino is the perfect tool with which to learn programming. The Arduino language, the Integrated Development Environment (IDE), is a dialect of the Processing language. This was created at the MIT Media Lab specifically for artists, musicians and non-technical users. This makes it one of the easiest languages to learn.

The programming languages C and Java fell in love and had a baby and called it Processing. If you are looking for one language to learn well, Processing would be a good choice. Everything you learn about Processing can be applied to the Arduino IDE and will jump start you into C or Java if you chose these directions.

Processing was modified to include control of the inputs and outputs of the Arduino microcontroller. The modified Processing language morphed into the language called Wiring, which then became the Arduino IDE.

## 2.2 Your first step

Before we begin our adventures with an Arduino, we have to get an Arduino and arrange our computer to talk to it. In this chapter, I will show you where you can purchase an Arduino for less than $4 and how to install the IDE and then communicate between your computer and the Arduino.

If everything goes smoothly, this could take about 5 minutes. If there are complications, I will walk you through figuring out the problem and fixing it. When your computer can successfully upload a blank sketch to your Arduino, you can advance to the next chapter.

Fasten your seatbelt, let's dive into the quick start.

## 2.3 The Quick Start

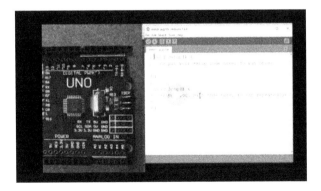

## 2.3 The Quick Start

*Click here to watch this video to see me walk through each of these steps quickly.*

If you just can't wait, this is the section to jump start your Arduino experience. I provide you the bare bones starting information. If anything you've read so far or read in this section is confusing, if it all sounds like jargon to you, if it's hard to follow, if you want a lot more details and step by step instructions, skip this section and go to the next section.

Here's what we will accomplish in this section:

1. *If you already have an Arduino, grab it. If you don't have one you can purchase one, along with a USB cable, from this web site I mention in my blog. If you want one right away, you can purchase it from Sparkfun using this link. If you have any problems, click here to jump to section 2.4.*

2. *Download the Arduino IDE from Arduino.cc. Then install it. If you have any problems with this step, click here to jump to section 2.5.*

3. *Launch the Arduino IDE for the first time. A blank sketch will open up. If you have any problems with this step, click here to jump to section 2.6.*

4. *Connect your Arduino board to a USB port using the appropriate USB cable. If you have any problems with this step, click here to jump to section 2.10.*

5. *Under Tools/boards, select the Uno board. Under Tools/port, select the COM port to which your Arduino is connected. If you have any problems with this step, click here to jump to section 2.11.*

6. *In the blank sketch that opens automatically, press the upload button and lights should momentarily flash on the Arduino board and you will see "Upload done" at the bottom of the sketch. If you have any problems with this step, click here to jump to section 2.13.*

7. *Congratulations! your computer can now successfully communicate with your Arduino. You have completed this chapter and can advance to the next or click to this section 2.15.*

If it's too fast, or you want more details, we do it all in the next few sections. If you are comfortable with this section, jump to Chapter 4 Getting to Blink.

## 2.4   Step 1: Purchase an Arduino for less than $4

Here is where we are in the process:

- ✓ **Get an Arduino.**
- ✓ *Download the Arduino IDE and install it.*
- ✓ *Launch the Arduino IDE.*
- ✓ *Plug in your Arduino board.*
- ✓ *Select the Uno board and the com port*
- ✓ *Upload the blank sketch*
- ✓ *Congratulations! your computer can now successfully communicate with your Arduino.*

To get us started, we need to purchase an Arduino. But, with more than a dozen different ones to choose from, which Arduino is the one for you? And of the hundreds of different places to buy one, where do you buy it? Let's get started.

In the Arduino ecology, there are more than 20 different Arduino boards. Figure 2.1 shows just a few of the options and their form factors.

*Figure 2.1. Examples of some of the Arduino form factors. The common element is that we use the same IDE to communicate with each of these boards in the Arduino family.*

With all of these options to choose from, which one should you start with? Each one of these has a different form factor, different strengths and a different set of weaknesses.

> *Unless there is a strong compelling reason otherwise, your first Arduino should be the Uno.*

The Uno is the simplest to use, the most robust, has the most open source code written for it, has the greatest number of shield add-on circuit boards that can plug into it, and is one of the lowest cost units.

I've used the Uno in my workshops for more than six years and in my own projects for almost twice as long. I have never had an Uno fail. My very hairy lab assistant, Maxwell, has ESD tested many of my experiments, and the Uno has passed every test. Figure 2.2 shows Maxwell hard at work on his latest test.

*Figure 2.2. Maxwell, hard at work ESD testing my latest Arduino project. He has not caused a failure yet.*

Generally, when I look to purchase an item that is a commodity device, that has many suppliers, I look at four different options from which to purchase the item:

- *Buy local if practical and affordable*
- *Buy from Amazon*
- *Buy from eBay*
- *Buy from an off-shore consolidating site like Aliexpress*

## 2.4 Step 1: Purchase an Arduino for less than $4

Since I live near Boulder, CO, I always look at purchasing items from Sparkfun. They are one of the leading suppliers of custom and standard parts for the maker community. You can also get parts in a few days, and since I live down the road from them, I can place an order on-line, and pick it up later that day.

They offer an incredibly valuable web site with lots of useful tutorials and project ideas. I often buy from Sparkfun for two reasons. First, I want to support a company that gives away a lot of valuable free content. In addition, they manufacture many of their own boards, like the Redboard, shown in Figure 2.3, and have an excellent reputation for providing quality parts.

*Figure 2.3. The Sparkfun Redboard, their brand of the Arduino Uno.*

When price is an important deciding factor, I first look on Amazon. Many commodity parts like resistors, LEDs, wires, cables and small tools can be found on Amazon. With prime membership, shipping for most items is free. Generally, the price for many items on Amazon is about half of what is found on other retail sites in the US.

Here is an example of a low-price Uno on Amazon, as shown in Figure 2.4.

*Figure 2.4. An example of an Arduino Uno on Amazon.*

When I am looking for the lowest price, the third place I go to purchase items is eBay. While it is often known as a bidding site, I rarely want to wait around for a bid to close. The "buy now" prices are usually so low, I just pay this price.

For some items, the price on eBay is half of the same item on Amazon. When shipping is free, eBay prices are generally a good buy compared with Amazon. When shipping is not free, eBay prices can be comparable to Amazon.

The biggest downside from buying low-cost electronics parts from eBay is that they generally ship from Asia and shipping can sometimes take 30-60 days. It's hard to be spontaneous when you have to wait weeks for your parts. When you need it quick, Sparkfun or Amazon is a better place.

An example of an Arduino Uno from eBay, as shown in Figure 2.5, can be found here.

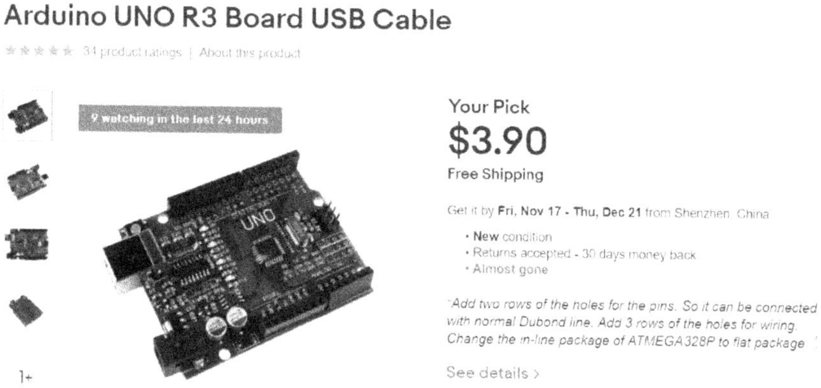

## 2.4 Step 1: Purchase an Arduino for less than $4

### Arduino UNO R3 Board USB Cable

Figure 2.5. An example of an Arduino on eBay.

Sometimes, there are so many choices from eBay, it's hard to find what you are looking for, or you may not be sure of the quality of the parts you get.

An alternative place to purchase low cost electronics parts is a consolidator, like AliExpress.com. Here, you can purchase an Arduino Uno for less than $3.50. This example is shown in Figure 2.6.

Figure 2.6. An example of an Arduino on Aliexpress for less than $3.50, complete with USB cable and free shipping.

These four examples illustrate the price range for the same item. From Sparkfun it is $19.95. From Amazon it is $7.99, from eBay, it is $3.90 and from Aliexpress, it is $3.25.

For less than $4, you can get the same Arduino Uno as listed for $20-$30 in other retail sites. This is complete with USB cable. The software to run the Uno is exactly the same and can be download for free. This is why the Uno is the best starting place for the first-time user.

There is one difference between the boards from these sources, related to the interface chip that goes between the USB and the micro controller. But, by following the directions I provide, it will have no impact.

The first step on your adventure is to purchase an Arduino. Wherever you get yours, be sure it comes with a USB cable. Most of them do.

Also, be sure it comes with the black header sockets already installed. These are highlighted in Figure 2.7. These header sockets are where you will connect external parts like LEDs, jumper cables and shields. You do not want to solder these header sockets on yourself but purchase the Arduino board with them already assembled.

*Figure 2.7. The Arduino board with one row of the header sockets identified.*

Ironically, we sometimes call these *pins*, when in fact, they are *sockets* or holes into which we stick pins.

What comes soldered to the Arduino board are header sockets that are soldered into holes in the Arduino circuit board. It's into these socket holes we insert jumper wires or component leads. But we still call these socket holes, pins.

Arduino boards from each of the four sources I mentioned above come with the header pins already assembled. Sometimes, an additional set of header pins are shipped with the Arduino, but these are for another purpose and we don't have to worry about them. Just put them aside.

When the Arduino is plugged into your computer, it will be powered over the USB port so no extra cords or power supplies will be needed. This is a cool feature. It means that you can plug the Arduino into a USB power socket or battery pack to power the Arduino after the sketch has been uploaded. This is really convenient.

While the starting price is under $4, and this is all you should have to pay, keep in mind the principle of the open source community. A great number of contributors give away their skills, their energy and their intellectual property, for free, so that you can benefit.

If you find value in what you have received for free or for a very low price, consider giving back to the maker community, either with your own contributions or by donation. This can be by donating to the Arduino.cc organization when you download the IDE, or by purchasing your parts from a local supplier, like Sparkfun, which contributes so much to the maker community.

If you are just starting out, if you are not hooked yet, if you are not sure this is an adventure upon which you want to embark, I think it's really okay to buy an Arduino for under $4 and not make a donation. Don't feel guilty, yet.

But once you're hooked, once you reach the point where you've been bitten by the bug and realize this is the coolest gadget since your set of blocks when you were three, you are now part of the maker community.

> *Contribute back to the community in your own way. Purchase parts from Sparkfun to support their great contributions, donate to the Arduino foundation, or pass your excitement to someone who doesn't know about these really cool devices.*

## 2.5 Step 2: Download the Arduino IDE

Here is where we are in the process:

- ✓ *Get an Arduino.*
- ✓ **Download the Arduino IDE and install it.**
- ✓ *Launch the Arduino IDE.*
- ✓ *Plug in your Arduino board.*
- ✓ *Select the Uno board and the com port*
- ✓ *Upload the blank sketch*
- ✓ *Congratulations! your computer can now successfully communicate with your Arduino.*

*Watch this video and I will walk you through downloading the IDE.*

While you wait for your Arduino to arrive, we're going to download the *Arduino IDE* (Integrated Development Environment) and check it out.

The IDE is the program we work in on our computer to write all the code that will run on the Arduino. In Arduino speak, the code is called a *sketch*. In the old days, we used to call this a *program*.

We create and write the sketches on our computer, using the IDE. Its purpose is to enable us to edit the sketches, store them and do some simple syntax checking. It will also convert, or *compile*, our high-level, easy to read sketch into the machine language the Arduino microcontroller chip understands.

We can't really do any serious debugging in the IDE as the IDE does not execute the sketch. The IDE editing environment runs on our computer, but the sketches run on the Arduino. We save the sketches on our computer so we can open them up later and do more editing.

Unfortunately, once a sketch is uploaded to an Arduino, there is no way for us to "suck it back out" of the Arduino and read it. This is another reason why we want to write, store and edit sketches in the IDE on our computer. It's the only place to which we have access.

To actually run the sketch, we need to move it over to the Arduino. We call the "moving over" step, *uploading* to the Arduino. Once the sketch is uploaded to the Arduino, it will stay there on the Arduino board in non-volatile memory. We can power off the board, and the sketch will still be there when we power up again.

The IDE is provided for free from Arduino.cc. This is a great site and should be on your goto list. It is one of the three web sites I recommend you bookmark and visit on a regular basis. These three web sites are:

Sparkfun

Adafruit

Arduino.cc

Initially, if you are not hooked yet on the Arduino, you should not feel any obligation to donate to the arduino.cc web site when you download the IDE.

> *However, once you are hooked, donating a little to the Arduino.cc site is one way of giving back to the Maker community.*

The IDE software can be downloaded from this location.

You are given two choices for running the IDE: in the cloud or on your computer. If you are just getting started, I recommend you downloaded IDE version that runs on your computer. This is generally more reliable and does not require an internet connection. An example of the landing page to download the IDE, as of this writing, is shown in Figure 2.8.

Download the Arduino IDE

*Figure 2.8. The landing page on the Arduino.cc site to download the IDE. This is for version 1.8.5. Your version will be probably be a higher rev.*

## 2.6 Step 3: Launch the IDE and a blank sketch is opened

Here is where we are in the process:

- ✓ *Get an Arduino.*
- ✓ *Download the Arduino IDE and install it.*
- ✓ **Launch the Arduino IDE.**
- ✓ *Plug in your Arduino board.*
- ✓ *Select the Uno board and the com port*
- ✓ *Upload the blank sketch*
- ✓ *Congratulations! your computer can now successfully communicate with your Arduino.*

---

> *Watch this video to see me launch the IDE, open a blank sketch and then explore the features.*

Once downloaded to your computer, install the IDE application by double clicking on the .exe program. Once installed, using all the default conditions, you will now see the Arduino program in your list of programs, and an icon to start the program, on your desktop.

I use a PC. On my PC, I like adding the program icons to my quick start menu so I can just click the link and the program starts. It's a personal preference.

When you open the Arduino program from the program menu, or double click on the icon on your desktop, or click once on the program link in your quick launch taskbar and run the program for the first time, it will open with a blank sketch.

We will compose the sketch on our computer, do some nominal error checking, and then upload it into the Atmel 328 chip that is on the Arduino board, all automatically with one mouse click.

Once the sketch is embedded on the board, it will run independently of being connected to the USB port. You just need to have the Arduino board connected to the USB port for power. Alternatively, once the sketch is loaded on the Arduino board, the board can be powered by a DC power source or even a battery and operate untethered from your computer.

Even if you unplug your Arduino, the sketch is still there and will run again as soon as you power it up.

When you run the IDE for the first time on your computer, a blank sketch opens up, such as shown in Figure 2.9.

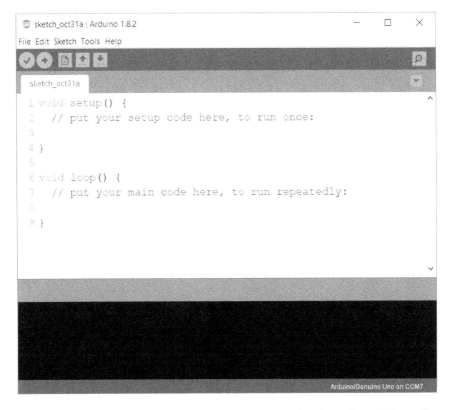

*Figure 2.9. An example of the blank sketch that opens up when the Arduino IDE runs for the first time or you select File/new.*

The Arduino language, often referred to as just the *IDE (Integrated Development Environment)*, is based on a combination of C, and Java Script which was transformed into a new language called *Processing*. This was created by the MIT Media lab specifically for artists and musicians, so it is one of the easiest languages to use. Check it out at www.Processing.org.

It is totally free and there are thousands of programs written in it that are all open source which you can download and try.

Processing was later modified to talk to the Atmel 328 series of micro-controllers and turned into a language called *Wiring*. In its final stage of evolution, special commands to access the features of the Arduino board were added to the language and it was re-named the *Arduino IDE*.

It still retains many of the features of C and Java Script. The investment you make in learning processing or the Arduino IDE will help you tremendously if you ever want to learn C or Java Script, or any other language.

If you have never done any programming, the Arduino IDE is a good starting place. And, if you like electronics or want to learn it, there is no better combination than the Arduino board and the IDE to get you started in physical computing.

## 2.7 Introducing the Syntax of a programming language

> your code that much denser and harder to debug.
>
> Curly brackets, {}, define *functions*. The third important syntax feature is the curly brackets, {...}. A very powerful way of breaking code up into manageable pieces is with *functions*.
>
> A *function* is a block of code that does something. The function's code is contained inside the curly brackets.
>
> We will see the opening bracket, {, at the start of the function and the closing bracket, }, as the last line of the function. Everything between them is run as the function. My buddy, Linz Craig, says the curly brackets are like the buns of a hamburger. The buns are

*Watch this video to see me walk through some of these features of the IDE.*

When we refer to the specific features of a programming language that relate to how it is written, we call it *syntax*. It's the grammar and spelling and punctuation rules of the particular programming language. Unfortunately, it's different in different languages.

> *Just like your 7th grade English teacher, the IDE is not very forgiving for simple spelling, grammar or punctuation mistakes.*

Syntax are the specific formatting details that have to be correct, or the IDE will complain and stop due to an error.

There are a few syntax features to always be aware of which I stumble on all the time:

Case sensitive. All commands, labels, and words are case sensitive. Every single little detail has to be correct.

Color coding. If you are typing a *built-in word* that is used for some command or constant, like *HIGH* or *digitalWrite*, it will appear in a special color. If, after you type the word, it is not a special color, you typed it in wrong. This is a big hint.

Comment lines. A line beginning with two forward slashes, "//" will not be read by the Arduino. You can write anything you want after the two forward slash marks. These are called comment lines. They are handy to place around so you can document what you are doing to leave yourself notes or breadcrumbs so others who may want to read your code can follow what you are doing.

> *Adding comments is a really good habit.*

In the blank sketch in Figure 2.9, lines 2 and 7 are comment lines.

## 2.7 Introducing the Syntax of a programming language

You can even leave a URL in a comment and have it hot linked to open the web site when it is clicked. I love this feature. I can put the URL address of where I got the code, or where there is additional part information right in the sketch that runs on the Arduino.

<u>All commands end with a semicolon</u>. All lines that are commands, ready to be executed, must end with a semicolon, ";". This is the most common error when writing a sketch- you forget to end the line with a ; . When the IDE checks your code, it will give you an error and often highlight the line in which you forgot to add the ;.

> *I tell my students, "computers should work for us, we should not work for computers."*

If the computer is smart enough to know to tell us we left the ; off, why doesn't it just add it to the line automatically?

It turns out that one way of typing a command that extends over multiple lines is to just add a return in the middle of the command and continue typing on the next line. The IDE uses the semicolon, ; to tell where the end of the command is. We just can't start a new command without ending the previous one with a ;.

In the blank sketch above, there are no command lines yet, so none of the lines end in a ;.

You can even have more than one command per line, if you end each command with a ;. This is not commonly done, as it makes your code that much denser and harder to debug.

<u>Curly brackets, {}, define *functions*</u>. The third important syntax feature is the curly brackets, {...}. A very powerful way of breaking code up into manageable pieces is with *functions*.

A *function* is a block of code that does something. The function's code is contained inside the curly brackets.

We will see the opening bracket, {, at the start of the function and the closing bracket, }, as the last line of the function. Everything between them is run as the function. My buddy, Linz Craig, says the curly brackets are like the buns of a hamburger. The buns are there to keep all the good stuff from falling out.

<u>Setup</u> and <u>loop</u> <u>functions in every sketch</u>. Every sketch has a minimum of two functions. The void setup (), beginning on line 1 in the blank sketch, is a special function. All the code between the beginning { and ending with the closing }, runs once only. This is a unique function in the Arduino IDE, and also processing.

The void loop () function, beginning on line 6 in the blank sketch, is also a special, unique function in the Arduino IDE. All the code between the { and the }, will run over and over again. It will loop.

These two sections of code are called *functions*. The word "*void*" in front of them describes the type of function. The word void means that there is no value returned by the function. The function is "*void*" of a returned value.

Once you have your Arduino, we will use this blank sketch to test our connection to the Arduino and any time we want to erase the current sketch, we will upload this blank sketch. It will overwrite any other sketch present and this blank sketch will run.

You can always create a new sketch by going to File/new and presto you have a new, blank sketch.

## 2.8 Summary of important vocabulary

Up to this point, we have introduced a few new terms which we will use over and over again. It's good to know what they mean so you are not confused. And, if you want to sound hip and part of the Arduino community, begin to use these terms when referring to the important elements.

IDE: Integrated development environment. This is the programming environment we use to write the code on our computer for the Arduino.

Arduino: family of boards, all compatible with the IDE

Microcontroller: the actual specialized processor chip on the board (Uno uses Atmel 328)

Sketch: the actual program or code that runs on the microcontroller which we write on our computer

Upload: the process of moving the sketch from the IDE on your PC to the microcontroller on the Arduino board

Shield: an add-on board that just plugs into the specific form factor of the pins on the Arduino board

Library: special drivers that allow us to use high level commands to drive special features in the shields

Functions: a small section of code that performs some special action

Syntax: the detailed structure of how the code is written and interpreted, like the grammar and spelling in a sentence.

## 2.9 Using the help menu to get started

When you look at the menu items at the very top of any sketch, like the new one in Figure 2.9, on the far right is the *Help* menu item.

The very first item you pull down under Help is *Getting Started*. This is shown in Figure 2.10. When you click on Getting Started, it takes you to the Getting Started page which was created and stored on your computer when you installed the IDE.

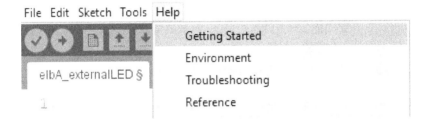

*Figure 2.10. Under the help menu is Getting Started.*

I've modified the steps in this video eBook from the ones on the Arduino.cc list slightly to make it a little easier and updated. We've already completed the first three steps:

- ✓ Get an Arduino.
- ✓ Download the Arduino IDE and install it.
- ✓ Launch the Arduino IDE.
- ✓ Plug in your Arduino board.
- ✓ Select the Uno board and the com port
- ✓ Upload the blank sketch

- ✓ *Congratulations! your computer can now successfully communicate with your Arduino.*

## 2.10 Step 4: Connecting your Arduino for the first time

Here is where we are in the process:

- ✓ *Get an Arduino.*
- ✓ *Download the Arduino IDE and install it.*
- ✓ *Launch the Arduino IDE.*
- ✓ **Plug in your Arduino board.**
- ✓ *Select the Uno board and the com port*
- ✓ *Upload the blank sketch*
- ✓ *Congratulations! your computer can now successfully communicate with your Arduino.*

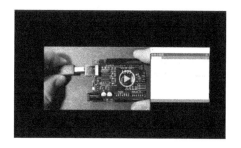

*Watch this video to see me walk connecting my Arduino board to my computer.*

When you install the Arduino IDE, the drivers are also installed so your computer can talk to the Arduino board. The IDE on your computer should be able to communicate to the Arduino board. After the IDE is installed, plug the Arduino board into the USB port of your computer with the cable.

Pay attention to three important clues that everything is working okay:

- *Listen for the "happy" sound of the driver loading*
- *Watch the power light turn on and stay on*
- *Watch for some flashing lights indicating communications between the computer and the Arduino*

Sometimes, depending on the Arduino, the on-board LED attached to pin 13 might stay on or blink.

Listen for the happy sound of a device driver being installed and now connected. You should see a few lights on the Arduino board light up. One light will be on continuously, the power light. Its location and color vary from board to board.

The second light that may be on may be flashing. One of the test programs that is run at the factory is to blink the on-board LED off and on. If your board shows a flashing light, this is a good sign.

Figure 2.11 shows my Arduino board with just the power indicator light on.

*Figure 2.11. My Arduino board plugged into the USB port of my computer showing the power light on.*

If the power light on your Arduino is not on, this indicates either a bad cable, a bad USB port, or a bad board. You should try alternately replacing each one of these components and re-testing for the power light.

If the LED power light is still not on, it may be a bad board. This is one reason to buy from a quality supplier like Sparkfun.

If the power light is on, you are ready to test the communication to the Arduino.

## 2.11 Step 5: Select the board and com port

Here is where we are in the process:

- ✓ Get an Arduino.
- ✓ Download the Arduino IDE and install it.
- ✓ Launch the Arduino IDE.
- ✓ Plug in your Arduino board.
- ✓ **Select the Uno board and the com port**
- ✓ Upload the blank sketch
- ✓ Congratulations! your computer can now successfully communicate with your Arduino.

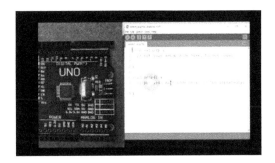

*Watch this video to see me walk through selecting the board and com port.*

As a final test of the Arduino and computer to verify they are communicating, we will move (*upload*) the blank sketch into the Arduino.

## 2.11 Step 5: Select the board and com port

There are two steps we have to complete before we can talk to the Arduino.

- *Select the correct board.*
- *Select the correct com port.*

To select the correct board, under the *Tools* menu, pull down and select *board*, and then the *Arduino/Genuino Uno* board from the list.

Your list may look different than my list, but you will have an Uno somewhere on your list. This menu item on my computer is shown in Figure 2.12.

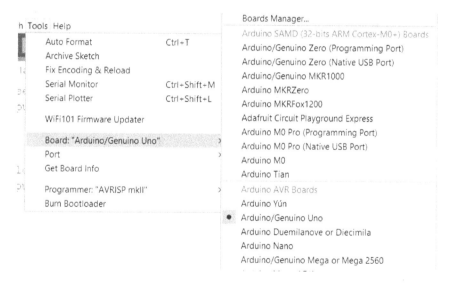

*Figure 2.12 Select Uno. In the pull-down menu under Tools/board select Uno.*

The next step is to select the port. Also under the *Tools* menu, pull down and select *Port*. You will see all the com ports available on your computer. An example on my computer is shown in Figure 2.13.

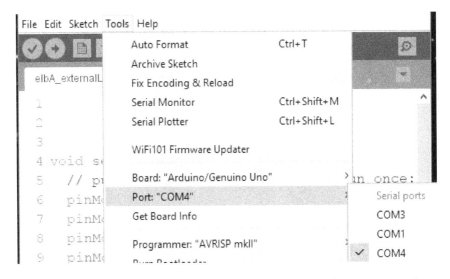

*Figure 2.13. Under Tools/Port, select the port to which the Arduino is connected.*

Which one is connected to the Arduino? Sometimes it's hard to tell. Sometimes you will see a clue. If one port name looks likely, try it.

Here is a trick you can use to identify to which com port the Arduino is really connected:

- *Disconnect the Arduino from the USB cable.*

- *Pull down the port menu and make a note of all the com ports listed. Obviously, NONE of these are connected to the Arduino.*

- *Now plug the Arduino back in and check the ports listed in the pull-down menu. You will see a new com port on the list*

- *The new port listed is the one to which the Arduino is connected.*

- *Select this one.*

On a Mac, the Arduino is probably connected to the port that is labeled as /dev/cu.usbserial-, or something like this. If it's not obvious, either try the trick above, or just try each port one at a time.

Be sure to select the specific port you want to try. Slide your mouse over a COM port and click the left mouse button to select the specific port, in my case, COM4. When your mouse moves over to the *Tools/Ports* menu item again, it should show the COM port you selected *check marked*, as shown Figure 2.13.

*Be sure there is a check mark next to the COM port.*

Now we are ready to do the final test: can we upload the blank sketch?

If you've been able to select the right board and the com port, skip the next section on troubleshooting. This section will walk you through how to test if you have the correct port by uploading the blank sketch.

## 2.12 Troubleshooting: no COM port or it's grayed out and nothing to select

With about 10% of my students, I encounter the problem that they can't select the COM port. It's grayed out or there is nothing listed.

The most common root cause is the driver that tells your computer how to talk to the Arduino board is not installed correctly on your computer. To fix this problem, there are four steps to try:

- *Unplug your Arduino, then plug it back in again. Is there a port to select now? If not, try the next trick.*
- *If you have multiple USB ports, try another port. Sometimes a specific USB port on your computer has an issue with the driver for the Arduino board. If a com port does no show up when you plug into a different USB port, try the next trick.*
- *Exit from your IDE program so it closes completely down, making sure there are no other sketches open. Then launch the IDE again and open a new blank sketch. Is there a port to select now? If not, try the next trick.*
- *Turn off your computer and reboot your computer. Runt eh Arduino IDE again and see if you can select the com port in a blank sketch.*

Each of these steps are trying to get your computer to recognize the driver for the Arduino board.

In very rare cases, none of these tricks fix the problem. This is the case in less than 1% of all the examples I've seen. If none of these tricks work, you may have to manually install the driver.

There are two different interface chips used on Arduinos. Many of the more expensive Arduinos, like the one from Arduino.cc and Sparkfun, use the FT232DRL chip. This requires the FTDI driver to talk to this chip.

Here is a great tutorial on how to manually install the drivers, from Sparkfun. You can download the driver from here.

The interface chip on the very low-cost boards, usually purchased from Asia, is the CH340g chip. You can download the driver for this chip from here.

To install this driver after you have downloaded the driver .exe file, unplug your Arduino and click on the exe in the driver folder. It will install the driver in your system. If you end up not using this chip on your Arduino board, no problem. It will have no impact on your system and do no harm.

If after installing both drivers, and restarting your computer, you still do not see a port to select, there is the possibility that your interface chip on your Arduino board is defective. Try another Arduino board.

In all of my Arduino experiences, a bad board occurs less than 0.5% of the time. It is not very likely, but with 1 million units out there, 5,000 boards will show this problem.

You may find a few of the suggestions on the Arduino.cc troubleshooting page, useful. Check them out.

Finally, if all else fails, try Googling your problem.

> *Remember, it is unlikely, but still a possible root cause, that your Arduino board is defective. If you do encounter a problem and nothing else works, buy another board.*

This is one of the good reasons to buy your board from a vendor like Sparkfun. They are there to help novices and experts alike.

## 2.13 Step 6. Upload a blank sketch

Here is where we are in the process:

- ✓ Get an Arduino.
- ✓ Download the Arduino IDE and install it.

- ✓ *Launch the Arduino IDE.*
- ✓ *Plug in your Arduino board.*
- ✓ *Select the Uno board and the com port*
- ✓ **Upload the blank sketch**
- ✓ *Congratulations! your computer can now successfully communicate with your Arduino.*

*Watch this video and I will walk you through this upload process and what the screens look like.*

On the line of icons below the menu items, underneath the word *Edit*, is a right-pointing arrow. This is the command button to upload the sketch over the USB port into the Arduino. This button is highlighted in Figure 2.14. Press this arrow button to upload the blank sketch into the Arduino.

## 2.13 Step 6. Upload a blank sketch

```
File Edit Sketch Tools Help
[✓] [→] [▣] [▲] [▼] Upload
sketch_nov01a
1 void setup() {
2   // put your setup code here, to run once:
3
4 }
5
6 void loop() {
7   // put your main code here, to run repeatedly:
8
9 }
```

*Figure 2.14. Find the upload button located under the Edit menu item.*

As soon as you press the upload arrow button, watch your board. You should see some lights flashing as the sketch is transmitted over the USB cable and received by the Arduino board.

At the bottom of the IDE screen on your computer is the status window. If you chose the correct COM port and everything worked, you should see the words, *Done Uploading* and some other information, as shown in Figure 2.15.

```
Done uploading.
Sketch uses 1994 bytes (6%) of program storage space. Max
Global variables use 184 bytes (8%) of dynamic memory, le

35                                          Arduino/Genuino Uno on COM4
```

*Figure 2.15. What appears at the bottom of the IDE indicating successful uploading.*

After the upload is successful, you should see the blinking light on the Arduino board turn off but the power light stay on. Sometimes the second LED on the board will stay on. As long as there are changes, the communications is working.

## 2.14 Troubleshooting- what if it didn't work? Enter Forensic Analysis

If you were not able to get to this final step, if the result is not what you expect, we switch our mindset to the *troubleshoot*, or *debug mode*. I like to call this *"forensic analysis"* as this sounds sophisticated. But, it's really what we are doing.

> *Just like a detective has to investigate who killed the victim, we need to find the guilty party. Who killed the project?*

The first step is to find the root cause. Then we fix the problem based on knowing the root cause.

We use the same tricks and techniques as a detective, looking for clues to point to potential suspects (root causes) and interrogate each suspect to see if they are the guilty party.

Like Captain Renault in Casablanca commanded, we will *"Round up the usual suspects."*

Based on years of teaching Arduino workshops to first timers, here are the usual suspects of possible root causes, or guilty suspects:

- *The cable is not plugged into a USB port*
- *It's a bad cable*
- *The COM button is not check marked*
- *Incorrect COM port selected*

## 2.14 Troubleshooting- what if it didn't work? Enter Forensic Analysis

- *A problem with the specific USB port*
- *The upload button was not the button pushed*
- *The correct driver was not installed on your computer*
- *The Arduino board is defective*

The first step in debugging is to check the cable. Is it plugged in between the Arduino board and the USB port of your computer? Try unplugging and plugging it in again. Do you hear any sounds from the computer recognizing you plugged a new device in?

Check that the COM port you want to try is check marked by pulling down the *Tools/Port* item again. Is the correct COM port checked? Try another COM port.

Try pushing the upload button again.

If the root cause is that the correct driver is not installed, try:

- *Unplug and re-plug in the Arduino board*
- *Make sure it is securely connected and both ends are well seated*
- *Completely close the IDE application and re-launch it*
- *Re-start your computer and try again*
- *Try a different USB port on your computer*
- *Re-install the driver and turn your computer off then on*

If none of these solutions work, try using another Arduino board.

## 2.15 Step 7: Congratulations!

Here is where we are in the process:

- ✓ Get an Arduino.
- ✓ Download the Arduino IDE and install it.
- ✓ Launch the Arduino IDE.
- ✓ Plug in your Arduino board.
- ✓ Select the Uno board and the com port
- ✓ Upload the blank sketch
- ✓ **Congratulations! your computer can now successfully communicate with your Arduino.**

If you were able to upload the blank sketch successfully, you are done with this Chapter. You have gotten through the biggest hurdle in getting ready to use the Arduino and explore its really cool features, which we begin in the next Chapter.

If you were successful after Step 1, thanks for coming along this journey with us. Maybe you learned a few tricks for the future.

You can now advance to the next chapter.

## 2.16 Summary of the new skills you have learned in this chapter

If you've gotten to this point, you should now be able to:

- Open up the Arduino IDE
- Open a blank sketch

## 2.16 Summary of the new skills you have learned in this chapter

- *Identify the setup() and loop() functions*
- *Select the correct board and com port*
- *Connect an Arduino to the USB port and identify the power on LED*
- *Upload a sketch to the Arduino and see the upload done sign in the IDE*

If you are not sure of any step, review the details in the sections above and try practicing them.

# Chapter 3. Getting to Blink

We will see how easy it is to control and manipulate the on-board LED. As simple as this component is, though, there are still a number of cool projects we can do just with the on-board LED.

## 3.1 What you need to know and what you will learn in this experiment

If you have some coding experience, it should take you about 10 minutes to complete the experiments in this chapter. If you have no coding experience, it may take you a little longer to pick up some of the tricks, methods and techniques I am going to show you.

You should browse the previous chapter if you have not already and verify that you are familiar with the terms and principles covered.

In this chapter, we introduce seven important features of the IDE:

- *Setting preferences*
- *Two functions: setup and loop*
- *Comments*
- *New Command: pinMode()*
- *Reference guide and in-context reference*
- *New Command: digitalWrite()*
- *New Command: delay()*

In the next chapter, we'll use these features to make our Arduino come alive.

If you don't want the details, you can take the quick splash and just get your Arduino blinking. Try it anyway, then check out all the details.

## 3.2 The Quick Splash

We are going to write our first sketch to turn the on-board LED off and on. Here is the complete sketch to write:

```
void setup() {
  pinMode(13, OUTPUT);
}
void loop() {
  digitalWrite(13, HIGH);
  delay(1000);
  digitalWrite(13, LOW);
  delay(1000);
}
```

If you want the super quick path, just copy the code out of this page of the PDF or eBook and paste it into a new blank sketch, overwriting whatever is in there. Then press upload. You will see the LED blink off and on.

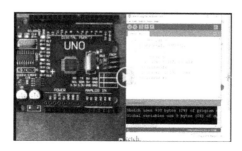

*<u>Watch this video and I walk you through creating and uploading the Blink sketch.</u>*

This will get you up and running with this sketch instantly. However, it's not something I recommend you do.

*The best way to learn to code is to develop the "muscle memory" of typing the code in so you remember each letter, each space and each character. If you really want to learn to code, type each example into a blank sketch. None of them are very long and you will learn the details faster.*

The more you type in code that works, the more you will get used to the format and the syntax which will make life so much easier when you are writing your own sketches.

Be sure to type this sketch into a blank sketch, exactly as it is shown. All the tiny details matter- especially the semicolon at the end of each line.

After you type in the sketch, there is a really cool tool to use to clean up your sketch.

Under the *Tools* menu, you will see at the very top, *autoformat*. After you type in this code, select *autoformat* and the code will be auto indented and look neat. The final code after autoformat is shown in Figure 3.1.

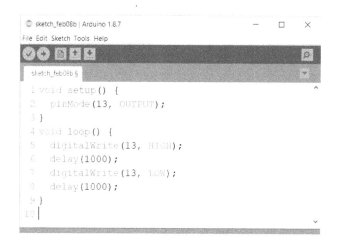

*Figure 3.1. An example of the blink sketch in the Arduino IDE after autoformat.*

This sketch, for obvious reasons, is called "Blink."

It is traditional, when you write your first sketch or program on any new computer, that you have the computer say (print) *"Hello World."*

This tradition was started by Brian Kernighan while he was at Bell Labs in 1972. A great description of the origin of this command can be found here.

When you print *"Hello World"* for the first time, you are connecting with this very rich tradition.

But the Arduino is not a *microprocessor* which is the brains in a computer. Rather, it is a *microcontroller*. It interfaces to the physical world and it is sometimes referred to as *physical computing*.

As such, rather than having your Arduino print "*Hello World*," Blink is a more physical version. It is *Hello World* for the Arduino.

*If you have gotten your Arduino to blink its LED
you have just connected with the rich tradition of
having your Arduino announce to the world,
Hello, I am here and ready for action.*

In the next chapter, I will show you how to make your Arduino come alive with a heartbeat.

**Try this experiment**: If you are comfortable with coding, here are some extra projects to try:

- *Modify the code to change the on time and the off time.*
- *How short an on-time can you see?*
- *How short an off time can you see?*
- *If the on time and off time is the same, how short an on and off time can you just barely perceive as still flashing. Going a little shorter makes the LED look like it is on continuously. This is the persistence of vision, or the flicker rate.*
- *As a stretch, try adding a few lines to make the Arduino's LED flash off and on like a heartbeat.*

Don't worry if you have no clue how to get started. We'll fill in the details.

## 3.3  Setting preferences and a good habit

When you launch the Arduino IDE for the first time, it comes up with its default settings. While this is an okay starting place, I find a few slight changes will make your life a bit easier. We change the default settings in the *preferences menu*.

### 3.3 Setting preferences and a good habit

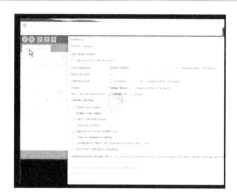

*Watch this video and I will walk you through setting the preferences menu.*

You will find the *preferences* menu under *File*. Click on *File* and select *preferences*, as shown in Figure 3.2.

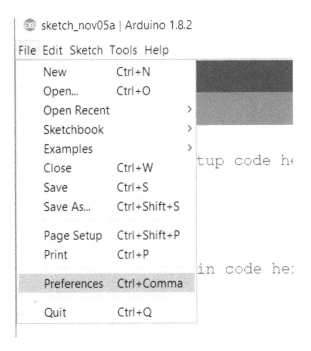

*Figure 3.2. Select the preferences menu under file to change some of the default conditions.*

When you select the preferences menu item, you will see the page to set all the preferences. What I like to use in my preferences is shown in Figure 3.3.

## 3.3 Setting preferences and a good habit

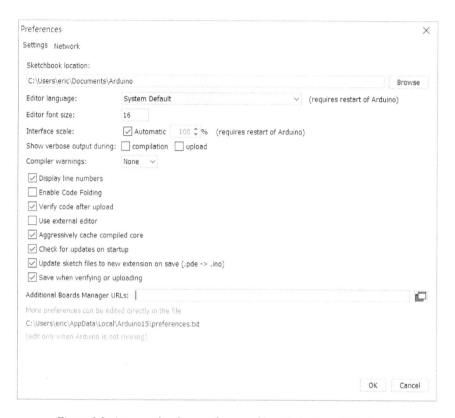

*Figure 3.3. An example of my preferences file with the items I like to use.*

I recommend making a few of these changes:

***Sketchbook location.*** When you save a sketch, it will go in this directory. It's really okay to use the default location. But, it's important to remember where it is. If you forget where the default directory location is, you can always come back and look it up here.

If you want a really easy place to find your sketches, I recommend creating a folder on your desktop with a name like *elbSketches*. Then select this as the folder to store new sketches.

> *As a special note, I try to make a habit to name all of my sketches with my initials at the beginning. This just makes it easier to tell at a glance these are the ones I created, as distinguished from being downloaded from the web.*

***Editor font size***. As a personal preference, I like using a larger font size than the default value. This just makes it easier to see on the screen. I like a setting of 16. When I present to a group and project on a screen, I use a font size as large as 24.

Experiment with this value to find the right combination for you between fitting a lot of lines on your screen and being large enough to easily read. It will depend on your eyes and the resolution of your monitor.

***Display line numbers***. When you select this box, which is what I recommend, line numbers adjacent to each line in your sketch will be automatically added to your sketch. I love this feature.

Line numbers makes it so much easier keeping track of your code, describing it to others and to instantly evaluate where you are in the sketch. This is a really cool feature and should be a default setting but is not. I recommend checking this box.

***Verify code after upload***. I recommend checking this box. All the Arduino IDE can do is check the *syntax* of your code before it translates it into the *machine language* the Arduino microcontroller understands. It cannot check the *logical flow* of your sketch. But syntax is a common source of errors. By checking this box, the IDE will verify that at least there are no syntax (grammatical) errors. There is no downside to checking this box.

***Save when verifying or uploading***. You can check the other boxes listed in my preferences file, or not. It's just a personal preference. But this last item is ***important***. I recommend checking it. This will force the IDE to save your sketch every time you check or upload your sketch.

*The first lesson you learn when you begin working with computers is, you cannot save your file too many times or too often.*

There is always the chance you've worked an hour on a sketch and then your cat walks over the keyboard and pushes the power off button and you lose all your work. I can't believe I have the only cat in the world that has shut down a computer

While it's a good habit to remember to save your work routinely, you can get the IDE to do it for you automatically, with this checked box. With this box checked, each time you do a syntax check of your code without uploading it, you will automatically save it. Each time you upload your code to the Arduino, you will automatically save it. Each time you click either of these boxes, your sketch will be saved automatically.

*DANGER Will Robinson*! There is one small problem with this setting. Sometimes we will be opening up and using a sketch we got from someone else, or one of the built-in examples. If this box is checked, and we make some changes to someone else's sketch and then upload it, the IDE will write over the original sketch. This just makes it harder to keep track what is in the original sketch and what is the stuff we added.

*To prevent this problem, we need to get in the habit of changing the name of any sketch we get from someone else as our first step.*

Whenever we open someone else's sketch, the very first step is to use *File/Save As* and give the file a new name. My own personal preference is to use my initials, "elb_", as the start of any new file name. This way I can tell at a glance which sketch is one I've modified and also where it came from.

*A special note about file names*. It is a good habit to NEVER use a space or funny character in a file name. You can use an underscore "_" to separate names, a dash "-"or use camel case, alternating lowerAndUpperCase letters to make file names more readable. Shorter names are better, but at least they should be self-documenting.

Just note that when we save a sketch, the IDE will ALWAYS put the sketch in its own folder with the same file name. The IDE does not like to see a "naked" sketch. The sketch always needs to be within its own folder.

If we do not specify a specific directory in which to place the sketch, it will go into the default directory listed in the preferences file.

As long as we do a *Save As* initially to any new sketch we opened from somewhere else, this auto save feature is a huge benefit. It's a good habit to get into.

Now the preferences are re-set and will be used every time you open the IDE on your computer. If you change computers, or re-install the IDE, you might want to check your preferences file.

## 3.4 A quick tour of the Arduino board

*Watch this video and I walk you through some of the features of an Arduino board and we will measure a HIGH and a LOW signal.*

Before we start getting into controlling the LED on the board, let's look at some of the features of the board.

I said at the beginning that the function of the Arduino is to measure input voltages and create output voltages.

Take a look at your board. Orient it so that the USB connector is at the top, as shown in Figure 3.4. This way the labels for each pin are easy to read.

*Figure 3.4. The Arduino board with the digital pins on the right and some analog pins on the lower left.*

There are two rows of header sockets on the board, some on the left-hand side and some on the right-hand side. The header sockets on the right side are all digital pins, labeled from 0 to 13.

The pins on the lower left side are analog pins, labeled A5 at the very bottom left, up to A0 and then a few other special voltage sockets above these.

We refer to each of these connections into the Arduino as "pins", even though they are holes, sockets into which we stick wires or other pins that connect on the board to the Arduino.

The digital pins on the right can be used to input digital signals or output digital signals in the form of 0 V or 5 V levels.

The analog pins on the left-hand side of the board, are really special. They can be used as digital pins, as digital inputs or outputs, and ALSO as analog input pins.

When the analog pins act as digital pins, they can input or output just the two states, a HIGH or a LOW, a logical level 1 or a 0. On our Arduino Uno board, the voltage output of a HIGH is 5 V. The voltage output for a LOW is 0 V.

In addition to the analog and digital pins that can output or input voltages, there are also a few on-board LEDs. In Figure 3.4, you can see one of the LEDs on. This is the power indicator. There are three other on-board LEDs, labeled, moving to the left, RX, TX, and L. These labels, and their position on the board will vary from board to board.

The LED labeled as L on this board is special. It is connected to digital pin 13. When digital pin 13 outputs a HIGH, the LED on the board will turn on.

Unfortunately, we cannot see or feel or smell voltage. We can measure it using an instrument such as a voltmeter, like the one shown in Figure 3.5. However, we can use the on-board LED, attached to digital pin 13, as an indicator to tell us when a 5 V signal is coming out of pin 13. The on-board LED is a convenient indicator at our disposal.

*Figure 3.5. A DMM measuring the output voltage of a digital pin set as a digital HIGH.*

In the rest of this Chapter, we'll see how to control digital pin 13 and the on-board LED. The techniques we introduce to control the on-board LED we will use in many future projects.

## 3.5 Command: pinMode(), the first command to bend the Arduino to our will.

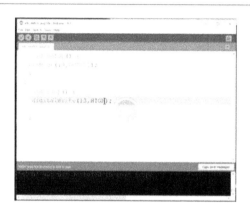

*Watch this video and I show you how to enter the pinMode and digitalWrite commands into a sketch and how to access help.*

The digital and analog pins each have multiple personalities. They can be inputs or outputs. Before we can use a pin, we have to tell it what role it is going to play: will it be an INPUT or an OUTPUT?

We use a command to tell the pin what role it will play. The command is pinMode. With this command, we will tell the Arduino which pin we want to control and what role it plays.

The syntax is pinMode (pin#, mode). The pin# is the number of the digital or analog pin, like 13 or 11 or 9. The mode tells the pin what role it will play, an OUTPUT or INPUT.

For example, to control pin 13 as an output, the command is

```
pinMode(13, OUTPUT);
```

In this command, the specific capitalization for every word is critical, as is the ";" at the end to tell the Arduino we are finished with the command.

The pinMode command has the M in caps. In the IDE, the command word, pinMode will be a special color and the word OUTPUT, will be a special color. If we typed the command correctly, the IDE will automatically color code the command. This is seen in the sketch displayed in Figure 3.1.

If we forget what the syntax is, we can always use the reference handbook which can be found under *Help/Reference* in the menu. This has a list of all the most common commands. Select on this page the pinMode command and you will see the syntax and a few examples for this or any command.

When we set up a pin for a specific function, we only want to do this once, so we place this command inside the

```
void setup() {
```

function.

If you have not already, open up a new, blank sketch and type in

```
void setup() {
   pinMode(13, OUTPUT);
}
```

In these examples, I think it is important for you to type in the commands rather than to open a stored or downloaded sketch. This is part of the *"muscle memory"* you should develop to understand each step of the sketch.

However, if you want, you can literally copy the text directly from this PDF or eBook and paste it into a new sketch, overwriting what is there. Your new code should run. Don't forget to do a *save as*.

***Here is a bit of a pro tip:*** When telling a digital pin to be an INPUT or OUTPUT, you just need to use the pin number. But, an analog pin can also be a digital INPUT or OUTPUT. Its numbers are A0 to A5.

The default assumption by the IDE is any number you use as a pin number is a digital pin. If you want an analog pin to be a digital INPUT or OUTPUT, use the pin number that includes the A, as A0, A1, A2, ... This way, you can make analog pin A4 a 5 V source using pinMode (A4, OUTPUT), and then turn that digital pin on later in the code.

*This allows us to use an analog pin as a +5 V and ground source to power sensors, if they do not require much current. We'll see this as a powerful trick for the future.*

## 3.6 New feature: the built-in Reference, where you can find all the commands

In Blink, we had to introduce just a few commands, pinMode, digitalWrite and delay(). These are just a few of the available commands.

There is a reason we refer to the elements that make up the code as a language. In a language, there are grammatical elements like nouns and verbs and punctuation. We combine these together into phrases and sentences. Then, we combine sentences into paragraphs and paragraphs into stories.

*A sketch is like a story. It has paragraphs, which are functions, and it has sentences, which are lines of code.*

Verbs are the action commands and nouns are objects that are acted on. Variables are personal pronouns that will be specific things or have general names. The syntax is the punctuation and grammatical rules.

It's hard to remember all the nouns and verbs, let alone the syntax for each one. Luckily, the dictionary with all the words we can use, is right at our fingertips.

Under the menu item *Help*, you will see *Reference*, as shown in Figure 3.6.

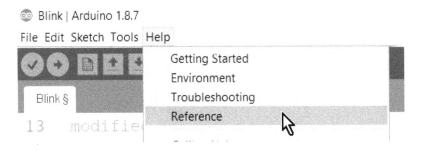

*Figure 3.6. Find the link to the reference page under the help meu.*

When you click this, it opens in your browser a page, stored on your computer when the IDE was installed, with a cheat sheet of all the most common verbs and nouns with their syntax.

### 3.6 New feature: the built-in Reference, where you can find all the commands

*This is a handy feature to remember, as everyone has to look up words in the dictionary at one time or another.*

There is another way to access the reference guide for a specific function in a sketch. If you have a sketch with some functions or commands and want to check the details of the syntax, or how the specific command works, you can access the reference guide for that function right from the sketch:

- *While the sketch is open, move your cursor over the word for which you want to find the description*
- *Click the word to place the cursor in the word.*
- *Then right mouse click and you bring up a contextual menu.*
- *Scroll down to the second from the bottom item,* **Find in Reference***.*
- *Select this item and the dictionary page for this command opens up.*

This item in the menu is shown in Figure 3.7.

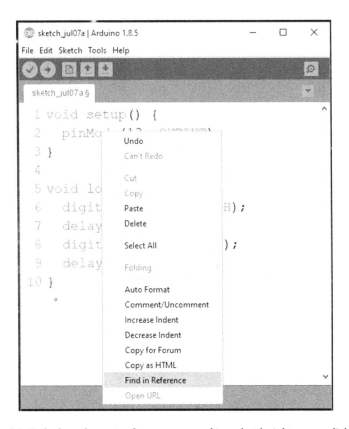

*Figure 3.7. To look up the syntax for any command in a sketch right mouse click on the word and select Find in Reference.*

## 3.7 New Command: digitalWrite(), Turning on or off output voltages

The next command we will learn is digitalWrite which will turn the selected pin on or off, making it a HIGH or a LOW.

The syntax is digitalWrite(pin#, HIGH or LOW);

If we want to turn on the output to pin 13 to be 5 V, it would be

digitalWrite(15, HIGH);

## 3.7 New Command: digitalWrite(), Turning on or off output voltages

> *When starting out, or writing complex code, we should get in the habit of trying out pieces of it along the way, just to check that we have it working correctly. This is a healthy process which makes debug dramatically easier.*

Let's create a sketch that just turns the on-board LED on. Think about what you would need in the sketch and try it yourself. If you get stuck, take a look at my sketch below.

```
void setup() {
  pinMode(13, OUTPUT);
  digitalWrite(13, HIGH);
}
void loop() {
}
```

All the commands are placed in the `setup()` function because it only needs to execute once. But, every sketch needs to have the `loop()` function, so I added it at the end.

While you can copy this code from this page and paste it into a blank sketch, overwriting everything in the sketch, I think it is useful to learn the "muscle memory" of writing the code to get a feel for the details. Type this into a blank sketch.

When you've written your sketch in the IDE and are ready to upload it to the Arduino and have it execute your code, there are generally three steps:

- ✓ *Make sure you have the correct file name you want to use for this sketch.*
- ✓ *Press the verify check mark button. This will check the syntax of your sketch.*

- ✓ *Press the upload button and watch the lights flash on the board and the compiling... uploading...uploaded words change on the bottom of the sketch.*

When you press upload, your LED should turn on and stay on.

You have just written your first sketch that controlled a voltage on your Arduino board and made it do something.

> *Welcome to the wonderful world of physical computing.*

Next, try turning the LED off, by adding a command line that changes the output value from `HIGH` to `LOW`.

So far, this is pretty dull and boring, but we will evolve this sketch. Combining these actions, we can write a sketch that will modulate the LED on and off and repeat this.

To repeat it automatically, we will place it in the void `loop` () function.

## 3.8 New command: delay()

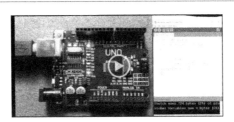

*Watch this video to see me walk through using the delay function to change the timing of Arduino operations.*

One way of controlling the speed of an operation is with the delay() function. This will tell the Arduino to sit there twiddling its thumbs, doing nothing, for the time, in milliseconds, inside the ().

We often abbreviate the word milliseconds as just msec. In one second, there are 1000 milliseconds. A value of 500 msec is half a second. In a later section, we'll do some experiments to estimate how long 200 msec or 500 msec is to try to get a feel for time intervals of a fraction of a second. The shortest time most people can respond to a stimulus, our reaction time, is about 200 msec.

We can use any value of delay we want, in msec, from 1 msec to 4,294,967,295 msec, which is about 50 days. Of course, during the time the delay() function is running, the Arduino is literally doing nothing, just sitting there twiddling its thumbs, waiting. But, if that's what you want it to do, the delay() command is perfect.

To make the LED blink at a rate we want, we would turn it on, wait some time and turn it off and then wait some more time.

Using our three commands, it would look like the following:

```
void setup() {
  pinMode(13, OUTPUT);
  digitalWrite(13, HIGH);
  delay(1000);
  digitalWrite(13, LOW);
  delay(1000);
}
```

As it is written here, inside the setup() function, between the first curly bracket { and the second curly bracket }, the commands will run through sequentially once and then stop.

## 3.9 New Code: use comments

Any line in a sketch that begins with two forward slashes, "//" is a comment line and is ignored. These lines do not take up extra memory and do not take any extra time in execution. They do not get turned into code loaded on the Arduino board.

This means you can sprinkle them throughout your code wherever they will help clarify what you are doing to someone reading it for the first time, or for you to review if you haven't seen your own code for a month.

In addition, you can add a URL in a comment line and it will be a live link. This is very useful when you want to document the details of the sensor you are using or point to a web site with more information about your sketch.

Whenever you declare a new variable, it's useful to add a comment after it with more information about the variable.

Whenever you add a new section of code, like a function, that has a special set of operations, it is useful to add a comment.

You can also use commented lines to separate sections of code to make it easier to read. Here are a few examples of using comments in my code:

```
long iTimeStart_msec;    // start time for the loop
// see www.HackingPhysics.com for more details

////////////////////////////
//global variables follow////////
////////////////////////////
```

### 3.9 New Code: use comments

```
float sensorLight_v;  // voltage calculated
int pinSensorLight;   // pin number for light sensor
long sensorLight_ADU; // ADU value of light sensor
//done with global variables//////////

////////////////////////////////////
```

If you want to comment out a large block of code, there is a special character to use.

At the beginning of the lines to comment, place a "/*" and at the end of the commented code, add a "*/". A long section of commented code would look like this:

```
/*
   all of these lines are commented
   and this line
   and this one
   until the last line here
*/
```

This long region of commented lines was used in the Blink sketch, available in the example sketches we'll see shortly.

*Here's a pro tip:* When you have a long sketch with a lot of pieces, you can comment out most of it to test out and de-bug the first few lines, then uncomment the next lines, then the next few lines, walking down through the code de-bugging and verifying the code as you go. This is one way to break up a large sketch into small pieces.

*Try this experiment.* Open up an example sketch from the *File/examples* list, like the very first one, *AnalogReadSerial*. Look at how the comment lines are used to document the sketch. Explore a few more sketches in the examples list.

## 3.10 My Sketch: Blink

We've assembled all the elements we need to get the on-board LED to blink in a simple off-on pattern.

If we want this process to repeat, we would place the turn on and off commands inside the void loop() function.

The complete sketch would look like this:

```
void setup() {
  pinMode(13, OUTPUT);
}
void loop() {
  digitalWrite(13, HIGH);
  delay(1000);
  digitalWrite(13, LOW);
  delay(1000);
}
```

Now we see that to change the on-time or off-time, it's just a matter of changing the 1000 inside the delay functions to any value of msec.

***Try this experiment***: Enter this code in a new sketch, change the on-time to a brief flash, upload and watch the Arduino's response.

*Congratulations, you have just written your first Blink sketch.*

## 3.11 Summary of the commands introduced so far

| Command | Description |
|---------|-------------|
|         |             |

## 3.11 Summary of the commands introduced so far

| | |
|---|---|
| `void setup(){`<br><br>`}` | This is a function that appears in EVERY sketch. Every command within the brackets will be executed just once |
| `void loop(){`<br><br>`}` | This is a function that appears in EVERY sketch. Every command within the brackets will execute over and over again. |
| `pinMode(13, OUTPUT);` | This command tells the Arduino that we are going to use a specific digital pin as an OUTPUT, as distinct from an INPUT. We use the pin number to identify the pin we want to setup. |
| `digitalWrite(13, HIGH);` | This command controls the output of a digital pin and makes it either a HIGH (logic 1, or 5 V) or a LOW (a logic 0 or a 0 V). Once executed, the pin value will be set to this value. |
| `delay(1000);` | This command tells the Arduino to sit there, twiddling its thumbs, doing nothing for a duration of milliseconds as listed inside the (). In this example, the time interval to wait is 1000 msec = 1 sec. |

| | |
|---|---|
| //<br>or<br>/*<br>*/ | Comments. Everything after these two forward slashes will be ignored by the Arduino sketch. We can use this add comments for our own benefit on any line.<br><br>For multiple lines add them between the /* and the */ lines. |

# Chapter 4. **It's alive**

As a graduation exercise to celebrate where we've arrived up the learning curve, we're going to make our Arduino come *alive*.

## 4.1 What you should know before you start this chapter

If you can write the Blink sketch and understand the commands, you know all you need to know.

These commands are:

```
void setup()
pinMode(13, OUTPUT);
void loop()
digitalWrite(13, HIGH);
```

## 4.2 Giving your Arduino a heartbeat

When I mentioned that Blink is the Arduino's version of *Hello World*, I lied a little bit.

Most of the world feels Blink plays this role. But I think there is a better application that announces to the world your Arduino is here and alive.

Instead of having the LED just blink off and on, I think it should flash like a heartbeat. It should flash in the rhythm of thump-thump....thump-thump....thump-thump....thump-thump.

I learned this trick from my buddy Linz Craig, another innovative Arduino educator expert and the founder of Questbotics, a company creating simple to use programmable robots for K-9 grades. A picture of Linz Craig with his small robotic cars is shown in Figure **4.1.**

Figure 4.1. Linz Craig, founder of Questbotics and the originator of the idea of making the Arduino LED pulse like a heartbeat, a much better Hello World, announcing your Arduino is ALIVE!

To make the Arduino's LED thump in rhythm to a heartbeat we need to figure out the shortest repeat pattern and put this in the loop function, so it repeats all by itself.

Think about the *algorithm* you would want to follow. These are the process steps you will tell your faithful assistant, the Arduino, to follow.

My first guess is something like this:

- *turn on for 100 msec*
- *turn off for 100 msec*
- *turn on for 100 msec*
- *turn off for 1000 msec*
- *"rinse and repeat"*

Take a stab at implementing this *algorithm*, or try your own. If it doesn't work the first time, or you don't like the timing, change the values until you think it is optimized.

If you get stuck, in the following section is my version of It's Alive. Try typing my sketch into a blank sketch.

## 4.3 My Sketch: It's Alive

```
//it's alive
void setup() {
  pinMode(13, OUTPUT);
}
void loop() {
  digitalWrite(13, HIGH);
  delay(100);
  digitalWrite(13, LOW);
  delay(100);
  digitalWrite(13, HIGH);
  delay(100);
  digitalWrite(13, LOW);
  delay(500);
}
```

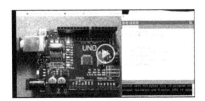

*If you get stuck, watch this video and I will walk you through bringing your Arduino to life.*

## 4.4 Printing "Hello World"

In the world of computing with microprocessors, we don't have access to LEDs we can flash. We just have keyboards and monitor screens. The tradition in microprocessor-based computers to announce to the world the computer is alive and ready for action is to print "Hello World" to the monitor screen.

Now that we have given our Arduino a heartbeat and announced it is alive, we can also follow the tradition and print "Hello World" to the screen. The process we use will be used over and over again to print lots of information to the screen.

All the information that comes back to our computer from the Arduino has to flow through the USB cable. When we uploaded the sketch to the Arduino, the code flowed through the USB cable from the computer to the Arduino. We will use it to also send information from the Arduino to the computer.

The USB cable is a serial data communications path. We refer to it as a serial link or a com port. For the Arduino to use it to send information back to the computer, we have to do three things:

1. Set up the serial link in the Arduino

2. Print information on the serial link

3. Have something on our computer to receive the information and display it.

These three steps are detailed in the next sections.

## 4.5 New Code: Serial.begin()

So far, the only way we have had to output information out of the Arduino is by flashing a light. We could change its brightens and flashing rate, but that was about it. There are other ways of getting information out. In this section we will look at printing information to the serial printer to appear on a terminal screen.

*Watch this video and I will walk you through the whole process of printing to the serial monitor.*

The first step is we set up the com link using the `Serial.begin` command. This command tells the Arduino to open the serial com link. This is only executed once, so it should be placed in the `void setup()` function.

The format for this command is very simple,

```
void setup() {
  Serial.begin(9600);
}
```

The word `Serial.` is a noun, in this example, specifically what we call an *object*. It is from such nouns that the family of languages referred to as, *Object Oriented Programming*, OOP, gets its name.

This *object* is the serial com port that links the Arduino to the computer and has a number of verb actions it can perform. In this case, we are using the verb command `begin`. The number we pass to the command, 9600, is the *baud rate* at which we want to send data over this serial com port.

The *baud rate* is basically the rate at which we send bits of information over the wires that connect between the Arduino and the computer.

A baud rate of 9600 is about 9600 bits of information per second. A bit of information is either a 0 or a 1. Generally, an ASCII character of text, like the letter a or B or number 4 or 9, takes 7 bits of information per character. This allows for encoding $2^7 = 128$ different characters. An extra bit is used in overhead, so it's basically 8 bits per character.

This means 9600 baud is about:

$$\text{characters per second} = \frac{9600 \text{ bits/sec}}{8 \text{ bits/character}} = 1200 \text{ characters per second} \qquad (1.1)$$

This is a rather slow rate of sending information over the com link. In many applications, it doesn't matter, but if we can do it reliably, faster is always better.

There is a limit to how fast we can send data over the com link before errors sneak in. This is set by the computer we use, the Arduino board and the cable connecting them. If we try to send data too fast, we may end up with a few errors. Once one error comes over, it may corrupt the rest of the data, so even one error can be bad.

This is why almost every sketch you see uses 9600 baud. What's special about 9600? Nothing really. It is just that at this slow rate, *every* Arduino and *every* computer using *every* cable you can find, should be able to send data over the com link without errors. In this sense 9600 is very *safe*.

If there is no compelling reason or need to go faster, 9600 is a good default value to use to make sure that everything is working. In the next section, we'll look at determining how fast we can send data before we get errors and then use this highest rate.

The command, `Serial.begin` (9600), sets up the com link. The next step is to send some data over it from the Arduino to the terminal screen.

> *Remember, if you want to see the reference page on this term, just click the cursor anywhere in the word in the sketch, right mouse-click and select the second from the last item, Find in Reference.*

Finding the reference page for the `Serial.begin` command is illustrated in Figure 4.2.

*Figure 4.2 An example of right mouse clicking on any command and opening up the reference page for that command.*

## 4.6 New Code: Serial.print()

Sending data using the `Serial.print` () command is easy. We are basically printing characters to the com link, either as letters or numbers or some other character.

Generally, there are two types of information we will want to send:

- ***text***, *which we call strings*
- ***numbers***, *which can be integers or floating point*

If we want to send text, we place it inside quotes. Any text inside the quotes will be printed exactly as it is typed in the sketch.

There are two print commands we will use:

- Serial.print("...")
- Serial.println("...")

The .print action will result in each word printed on the same line, one next to the other. This means everything goes on the same line.

The .println action will result in the item being printed, followed by a line feed- a carriage return to the next line.

All those items you want on the same line should be printed with .print. The last item on the same line should be printed with .println and the cursor will move to the next line.

It is sort of traditional, when we are sending data over the com link for the first time, to print, *Hello World*. This announces to the world that our sketch is alive. The command would look like this:

```
void setup() {
  Serial.begin(9600);
}
void loop() {
  Serial.println("Hello World");
}
```

This will print the words "Hello World" over and over again, each on a new line.

*One word of caution. The Arduino IDE only understands straight quotes. If we copy the sketch from a word processor, or some other document, sometimes, the document will use "smart" quotes,*

> *that are curly. If we copy and paste these into the IDE, we will get an error. We have to manually go into the sketch in the IDE and replace the curly quotes with straight quotes. This is another good reason to type the commands in to build muscle memory.*

If we wanted the `Serial.print` command to execute once, we'd place this command in the `void setup` () function. If we wanted it to loop forever printing over and over again on different lines, we would place it in the `void loop` () function.

## 4.7 New Feature: The serial monitor

To actually see the data, we need to open up the serial monitor from the Arduino IDE. This is the third step.

After the sketch has been uploaded and is running on the Arduino, it begins sending the data over the com port to the computer, whether we look at the data or not. To see the data on the computer, we have to open up a *terminal emulator* application on the computer to read the data coming in on the com port.

Luckily, we have a terminal emulator built right into the IDE. It is called the Serial Monitor. To turn it on, under the tools menu, select *Serial monitor*, as shown in Figure 4.3.

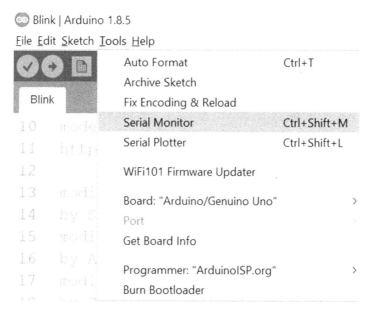

*Figure 4.3 Under tools, select the Serial monitor and a window will open to show the data received over the com link.*

Selecting the serial monitor menu item does two things. It opens up the terminal screen to see the data printed from the com port and received by the computer, and it also sends a reset command to the Arduino. This tells the Arduino to start its sketch over from the beginning.

> *It is important to remember that each time the serial monitor is opened by selecting it from this pull-down menu, it sends a reset command to the Arduino and the Arduino will stop what it is doing and go to the beginning of its sketch and start all over again. This will happen every time the serial monitor is opened from the tools menu.*

Once the serial monitor is open, the most important step is to make sure that it is set to receive data at the same rate as we set up the Arduino to send data.

In the lower right of the serial monitor screen is a small pull-down menu with all the data rate options. If the 9600 is not selected, pull down the menu and select it. Some of the values you can select from are shown in Figure 4.4.

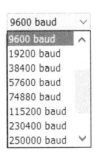

*Figure 4.4 The pull-down menu showing some of the baud rates possible. Make sure you select the same value in the serial monitor as you wrote in the Serial.begin() command of the sketch.*

Since we hard coded 9600 as the baud rate for the Arduino's serial monitor communications, be sure to select this in the serial monitor.

When the Serial Monitor is opened and the 9600 baud rate is selected in the lower right of the screen, the Arduino will be reset and you should immediately see Hello World printed over and over again scrolling down the screen. This output is shown in Figure 4.5.

## 4.7 New Feature: The serial monitor

*Figure 4.5 The serial print sketch and output from the serial monitor showing Hello World forever.*

When you get to this point, you are able to communicate from the Arduino back to the computer. This is a powerful feature we will use over and over again.

Of course, just because I told you to use this value in the serial monitor, don't feel constrained to not experiment. You won't break anything.

***Try this experiment***. Select the baud rate in the serial monitor as a value other than what you selected in the Serial.begin() command. You should see gibberish on the serial monitor.

It's only when using the same rate will you get clean data transfer.

If we want to add structure to the printed line, we can always add additional print statements with any features we need, like:

```
void setup() {
  Serial.begin(9600);
  Serial.print("Hello World");
  Serial.print(",   ");
  Serial.println("last line.");
}
```

Notice that every time you close the serial monitor and open it again, the TX and RX lights on the Arduino flash and it starts over again and prints from the beginning.

There is another way of forcing the Arduino to start over again from the beginning. Every Arduino has a *reset button*. The reset button on the Arduino Uno board I am using for these examples is located in the upper right corner in Figure 4.6.

*Figure 4.6 The reset button will force the Arduino to start its sketch from the beginning.*

Pushing this button will also force the Arduino to reset and start over again from the beginning of its sketch.

## 4.8 How fast can you print to the serial monitor?

Most sketches you will see in books or web sites use a baud rate of 9600. This is safe, dull and boring. If your application does not require very fast data transfer, it's perfectly fine. But what if we want to send data faster? Just how fast can we go?

*Watch this video where I walk you through testing the highest baud rate between your Arduino and computer over the USB com link.*

While we can type any value we want in the `Serial.begin()` command in a sketch, only some values can be selected in the serial monitor. These are the only values you should use so that you can match the `Serial.begin` value with the pull-down menu in the serial monitor.

Here are the values we can use for the baud rate. Notice that all the other values are multiples of 9600:

9600 (1x)

19200 (2x)

38400 (4x)

57600 (6x)

74880 (8x)

115200 (12x)

## 4.8 How fast can you print to the serial monitor?

230400 (24x)

250000 (26x)

500000 (52x)

1000000 (104x)

2000000 (208x)

In this section, we're going to experiment to see how fast we can send data before we get gibberish.

*How do you think you should do test this?*

***Try this experiment***: Take a few minutes to think about how you could find the highest communications speed before an error creeps in.

Here is the simple method I used. The sketch we just developed can be used to test the communications link. Here it is:

```
void setup() {
  Serial.begin(9600);
}
void loop() {
  Serial.println("Hello World");
}
```

That's it. That's the entire sketch. You can type this into a new, blank sketch. After it uploads, open up the serial monitor from the *Tools/Serial monitor* menu.

Just remember, it's the Arduino that is talking through the USB cable to the serial monitor.

Select the baud rate at the bottom of the serial monitor screen to match what you set in the `Serial.begin` () line.

Using my cheap Arduino Uno and the cheap cable that came with it, I never got an error printing on the serial monitor using the highest baud rate setting in the serial monitor, 2000000. This is really cool!

> *Knowing that I can get error free transmission at a baud rate of 2000000, this has become my new default baud rate to use. You will see it in many future sketches.*

***Try this experiment***: How fast can you use your com link? Not all interface chips that translate between the microcontroller and the USB port are capable of transmitting at this speed.

The result of this experiment is that I can routinely get reliable communications using the fastest baud rate of 2000000. If this is the case, why wouldn't we always use this fastest rate?

> *In all future sketches we will use a baud rate for the serial comm link of 2000000 baud. This will give us the fastest communications rate possible.*

## 4.9   Summary of the commands introduced so far

| Command | Description |
| --- | --- |
| void setup (){<br>} | This is a function that appears in EVERY sketch. Every command within the brackets will be executed just once |

## 4.9 Summary of the commands introduced so far

| | |
|---|---|
| `void loop(){` <br> `}` | This is a function that appears in EVERY sketch. Every command within the brackets will execute over and over again. |
| `pinMode(13, OUTPUT);` | This command tells the Arduino that we are going to use a specific digital pin as an `OUTPUT`, as distinct from an `INPUT`. We use the pin number to identify the pin we want to setup. |
| `digitalWrite(13, HIGH);` | This command controls the output of a digital pin and makes it either a `HIGH` (logic 1, or 5 V) or a `LOW` (a logic 0 or a 0 V). Once executed, the pin value will be set to this value. |
| `delay(1000);` | This command tells the Arduino to sit there, twiddling its thumbs, doing nothing for a duration of milliseconds as listed inside the (). In this example, the time interval to wait is 1000 msec = 1 sec. |
| `//` <br> or <br> `/*` <br> `*/` | Comments. Everything after these two forward slashes will be ignored by the Arduino sketch. We can use this add comments for our own benefit on any line. <br><br> For multiple lines add them between the /* and the */ lines. |

| | |
|---|---|
| `Serial.begin(9600);` | This command opens up the serial com link with a baud rated inside the (). In this example the baud rate is 9600 baud. The name `Serial` is an object, the serial link. The `.begin` is a verb, telling the object to get set up. |
| `Serial.print("Hello World");` | This command will print characters to the serial communications channel. Everything inside the () will be printed. To print a string, enclose the characters within quotes. After printing, the cursor will be left on the same line. |
| `Serial.println(" ");` | Does the same thing as the `Serial.print` command, but moves the cursor to the next line after printing the characters. This starts the next printed content on a new line. |
| Serial monitor | The terminal emulator that you can print to. This is the primary way of displaying information from the Arduino. |

# Chapter 5. Modulating the On-Board LED and Persistence of Vision

In our blink sketch, we manipulated the on-board LED in the most basic and simple way: we turned it off and on with the same delay.

In our *It's Alive* sketch, we manipulated the on and off times and changed the pattern.

This demonstrated the basic structure of a sketch to control the LED to make any pattern we want.

In this experiment, we will write a sketch which will turn the LED off and on to generate a simple pulse train. This is one of the basic building block signals we will use over and over again.

## 5.1 What you need to know and what you will learn in this experiment

You should be comfortable using the following commands:

```
void setup()
pinMode(13, OUTPUT);
void loop() {
digitalWrite(13, HIGH);
delay(100);
Serial.begin(9600);
Serial.print("Hello World");
Serial.println(" ");
//comments
```

In this chapter, we will introduce a valuable feature: variables, their care and feeding. Every sketch we use going forward will use variables. This will introduce you to the abstract thinking of algebra and the essence of programming.

## 5.2 The basic pulse train pattern

A pulse train is a repeating pattern of on and offs. We can describe this pattern with a few *figures of merit*. A figure of merit is a number that characterizes a behavior. It is based on using an ideal pattern as a template with the figures of merit describing specific features. We compare our signal to the template and what value of figure of merit makes the ideal signal match our actual signal.

For example, to describe a pulse train of on's and off's, we could use:

- *On-time*
- *Off-time*
- *Off-voltage (low)*
- *On-voltage (high)*

The generic structure of this ideal pulse train, identifying the figures of merit, is shown in Figure 5.1.

## 5.2 The basic pulse train pattern

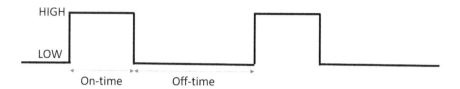

*Figure 5.1. Example of a simple pulse train of repeating on-time, off-time, on-time, off-time pulses.*

This is exactly the structure of the Blink sketch. We can literally use it and just adjust the on-time and the off-time. Here is how we would use it:

```
void setup() {
  pinMode(13, OUTPUT);
}
void loop() {
  digitalWrite(13, HIGH);
  delay(1000); // on time
  digitalWrite(13, LOW);
  delay(1000); //off time
}
```

This is a great opportunity for you to get your sense of short time intervals calibrated.

***Try this experiment***: Set the on-time and off-time to be 1000 msec and get familiar with what this time interval feels like. This will give you a calibrated 1 second interval (the on-time) with which to get comfortable.

Then, try this with a 500 msec on-time. Can you train your eye to judge how long ½ a second is?

## 5.3 What is the flicker rate?

*Watch this video and I will walk you through this first experiment.*

As written, the sketch in the last section will modulate the LED with the same on and off time. Using this pattern, we can explore the questions of how fast we can modulate the LED so that it appears to be continuously on.

As we decrease the off and on time, we reach a point where the LED looks like it is no longer blinking, but on continuously. We call this rate the flicker rate.

We already introduced one set of figures of merit that describe this pulse train, the on and off times. We can describe this behavior with another set of parameters or figures of merit.

This is a *periodic* behavior. The same pattern of on and off lights happens over and over again. We call one complete pattern a *cycle*.

One term that describes a property of a cycle is the time it takes to complete. This is called the *period*. If the on-time is 0.5 seconds and the off-time is 0.5 seconds, the period is the total time for one cycle or 1 second per cycle.

An alternative way of describing the properties of a cycle is its *frequency*, how many cycles are completed per second. The frequency of a periodic event is also referred to as a *rate* or *speed*. These terms are a little ambiguous. The *frequency* of the pattern is a better term to always use.

If we know the period per cycle, the frequency is how many cycles occur per second. The frequency is the inverse of the period:

$$\text{Frequency} = \frac{1}{\text{Period}}$$

If the period uses units of seconds per cycle, the units of frequency are cycles per second. For historical reasons, we call cycles per second, Hertz, abbreviated Hz.

The flicker rate, or flicker frequency, is the lowest frequency at which the LED appears to be on continuously. At a slightly lower frequency, the LED will appear to noticeably flicker. A higher frequency than the flicker rate and the LED will appear to be on continuously.

To calculate the flicker rate, first decrease the off and on time until the LED appears to not be blinking. The total time, the off-time + the on-time = the *period* of the pulse train. The flicker frequency is

$$\text{Flicker Frequency [Hz]} = \frac{1}{\text{Period [sec]}} = \frac{1}{\text{on-time} + \text{off-time}}$$

For example, when I adjusted the on and off time, I found that when the on and off time was each 20 msec, the LED appeared to be on continuously. The period was 40 msec = 0.04 sec and the flicker frequency was 1/0.04 sec = 25 Hz.

This is close to the value of 24 frames per second (fps), which is the frame rate used in movies. Why 24 fps? A slower rate and there is perception of individual frames. A faster rate and you use up more film. To reduce the cost of the film, the absolutely lowest rate that is still flicker free was selected. For historical reason, 24 fps was adopted. You can read about it here.

Here is my sketch to drive the on-board LED at 25 Hz:

```
void setup() {
  pinMode(13, OUTPUT);
}
void loop() {
  digitalWrite(13, HIGH);
  delay(20); // on time
  digitalWrite(13, LOW);
  delay(20); //off time
}
```

If we pulse the LED at a rate slower than 25 Hz, we can see the LED is flashing. If it is faster than 25 Hz, the flashes blur together in the after image in our eye and it looks continuously on.

***Try this experiment***: Use the Blink sketch to measure the flicker rate for you and your friends.

In many cars, the red lights used for brake lights or the back lights when the headlights are on, use red LEDs. These are often flashing off and on, but at a rate faster than most peoples' flicker rate.

The way to tell if a car's lights are actually flashing is to move your eyes rapidly around. Your peripheral vision will pick up the lights as streaks. If the light is really being flashed on and off, the streaks will have dashes in them.

When I drive at night, I sometimes move my eyes rapidly side to side when I look at a car's back lights. When I see short dashes in the afterimage, I can tell the car lights are LEDs and they are being modulated.

## 5.4 Changing the apparent brightness

If we use a total period that is shorter than the flicker period, the LED will appear to be on continuously. The persistence of vision in our eyes keeps a little of the image of the on-LED in our vision even when it is off. It's like our vision has some memory to it. The LED has to be off for some period of time for us to forget it was ever on and see it as an off.

*To pulse the LED so it looks continuously on, we need to keep the sum of the on and off times shorter than 40 msec.*

We can change the apparent brightness of the LED if we make the fraction of time the LED is on shorter or longer.

If we use an on-time of 40 msec, and off-time of 0 msec, the LED will always be on and it will look the brightest.

If the on-time is 20 msec and the off-time 20 msec, the brightness will look mid-range.

If the on-time is 1 msec and the off-time is 39 msec, the LED will look very dim.

**Try these experiments:**

1. Modulate the LED at the three different brightness levels and compare their relative brightness. Since the fraction of time the LED is on is very precisely controlled, the relative brightness is very linear.

2. *This is a chance to calibrate your eye. Does the sensation of brightness match what you set?*

3. *As you move your eyes around quickly, do you see the length of the LED streak change length with the on-time?*

The *apparent* brightness is really related to the *fraction* of the *whole time interval* in which the LED is on. The larger the fraction of time the LED is on in one cycle, the brighter it will look.

We call the fraction of time the LED is on, the *duty cycle*. When the duty cycle is 100%, the LED will be the brightest. When the duty cycle is 50%, it will be dimmer, and when the duty cycle is 2.5%, with the on-time 1 msec out of a total interval of 40 msec, (1/40 = 2.5%), it will be dimmest, before it is finally off.

**Try different duty cycles and see if you can calibrate your sensation of brightness with duty cycle.**

## 5.5 New Features: Introducing variables and variable types

So far, when we wanted to identify a specific pin, or we wanted a specific delay time, we typed the specific number into the sketch. We call this *hard coding* the number. Every time we wanted to change that number, we had to go in and type a new number. This worked but it was very rigid and awkward. By using *variables*, we dramatically simplify coding.

Variables are names we give to parameters that will hold the value of a number or character. They are a symbolic placeholder for some set of data. We can use the variable name to replace the number or characters in any calculation or operation.

## 5.5 New Features: Introducing variables and variable types

In computer-speak jargon, the *variable* name is the name of the location in memory in which we store the value of the data. Whenever we use the variable name, it points to the data stored in that location. This is illustrated in Figure 5.2.

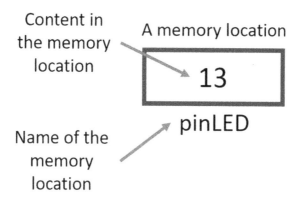

*Figure 5.2 How to think of a variable as a location in memory that stores a number.*

In the Blink sketch, pin 13 is the pin to which the on-board LED is connected. We *hard-coded* the number 13 by typing it explicitly in the digitalWrite command. We could have given the on-board pin number a variable name, or memory location label, such as the made-up name, pinLED.

We tell the sketch *pinLED* is a variable and create the variable location by declaring it at the very beginning of the sketch using a special syntax. At the same time we declare the variable, which creates it, we can also place in its memory location the number 13.

At any future time, we could refer to the pin number by this variable name and write the command, for example, as

```
digitalWrite (pinLED, HIGH);
```

In every command in which we need to address pin 13, we could use its variable name, pinLED. Whatever command sees the variable name pinLED will look up in its memory location the value stored there and use this value. If we later decide to use pin 12 instead, we just change the value of the number stored in the pinLED variable name location and pin 12 is used everywhere.

Every variable has three elements:

- *a type*
- *a name*
- *a value*

The *name* and *value* we described above. There are multiple types of variables based on what sort of data is stored in its location, such as: *int, long, float, Boolean, string, char,* and *arrays.*

In most of the sketches we will write, we will use two types of numbers: *integers* (a whole number with a plus or minus sign) like 448, 12, -94 or -5, and numbers with a decimal point, we call *floating point* numbers, like 4.81 or -2456.12.

The difference between an integer and a floating point number is that a floating point number has a decimal point, but an integer does not.

When we create a *variable* we want to use to represent a number, we have to decide, when we create it, if we want the number to be an *integer* or a *floating-point* number. The *number type* influences the sort of math we do with it and how the variable can be used.

For example, if we want to use the variable as a pin number, it must be an *integer* type. We create integer type numbers using the simple command

```
int pinLED;
```

## 5.5 New Features: Introducing variables and variable types

Some functions, like delay() will only use integers. However, if you type a delay of 145.57 inside the (), the delay function will turn it into an *integer* by truncating everything after the decimal point.

If we want to use the variable in a delay function as a delay of some number of milliseconds, it is a good habit to use a number that is an *integer* type. This way, we know exactly what delay value will be used.

The value of a number created from reading an analog pin is an *integer* type.

An *integer* type number has no decimal point. It is only a whole number. If we try to assign a number like 15.3 to a variable we created as an integer variable, only the whole part of the number, 15, will be stored in the variable location.

*This means that when we do some algebra like 5/2 and assign the value to a variable that is an integer type, only the integer part of the answer will be stored.*

Sometimes, this is a useful feature, such as when we want an answer as a whole number. Sometimes this results in a number we did not want, like when we are taking an average and the average is less than 1. If assigned to an integer variable, its value would be 0.

There are some limitations to the variable type "int". A variable defined as an *int* can only have a value between -32,768 to 32,767.

This is not a very large range. If we use an `int` variable to count milliseconds, the longest amount of time we can count is only 32.7 seconds.

If the value of our variable is already 32,767 and we add 1 to it, it rolls over to -32768 and counts up from there. This could be very inconvenient if we don't plan for it.

There will be many situations in which we want larger integer values. In this case, there is a different type of integer we can create, called a *long*. Instead of using the *int* command to declare an integer, we use the long command and declare a `long` integer. The command looks like:

```
long iCount;
```

A `long` variable can be between -2,147,483,648 to 2,147,483,647. If we use a `long` type variable to count milliseconds, we could count as many as 2 million seconds, which is about 1 ½ months.

> *You can never go wrong creating all your integer variable types as long. It will take up a little bit more memory space, but if memory is ever a problem, there are other microcontrollers you can switch over to where memory is not a limitation. They will not be the lowest cost, but they will still be low cost, and higher performance.*

We create a variable in the beginning of a sketch before the void setup() function. The syntax is the type of variable and the variable name. We can include an initial value if we want. Here are a few examples:

```
int iCounter=0;
long iTime_msec=17;
float V_tempSensor_V;
```

## 5.5 New Features: Introducing variables and variable types

### Try these experiments:

1. Any number assigned to an integer will always be in the format of an integer. What will int i1= 3.1415 be? Try creating the variable, assigning it this value and printing the variable to the serial monitor.

2. The largest integer you can store is 32,767. if you try to make an integer with a value of 32,768, it will roll over to the very beginning and start up. What value is i1=32768 or higher?

3. If you declare the integer as a long, you can count larger integers. What is long i1=40000, compared to int i2=40000?

4. When we declare a variable at the beginning of a sketch, we can:

    a. just declare the memory space, int i3

    b. declare the memory space and set a value to the variable, int i3 = 12

    c. declare the memory space, and perform a simple calculation using other, previously defined variables. long i4 = i2 + i1

5. What is int i2=14+13.6? What is long i9=329000+29?

Remember, you can print a variable's value to the serial monitor using just a few lines of code, like:

```
int i1=3456;

void setup() {
```

```
  Serial.begin(2000000);
}
void loop() {
  Serial.println(i1);
}
```

*Floating-point* type numbers are numbers with a decimal point. A *floating-point* type variable cannot be used as a *counter* in an *if* statement or as a *pin* number or in the delay function.

However, the *floating-point* type number is incredibly useful when describing a voltage, or a temperature.

> *Generally, if there is no compelling reason why a number should be an integer, or we think that when we use the variable, it may represent a number with a decimal point, the variable type should be assigned float.*

To use a variable, we have to first create it, or in computer jargon, *declare* it, by defining what type of number we want it to be, *integer* or *floating point*. It's a good habit to declare the variable at the beginning of the sketch before the setup() function. The command lines to create an integer or a floating-point variable, labeled as pinLED and sensorVoltage are:

```
int pinLED;
float sensorVoltage;
```

The command to declare an integer variable is int.

This command allocates a little space in memory, labeled pinLED, that will store an *integer type* number. At some point in the sketch we need to fill this memory location with the value of the number we want. We can both declare the variable and assign it a value all in the same line, such as with

```
int pinLED = 13;
```

In most cases, we place these lines of code which declare the variable and allocate the memory space, at the beginning of the sketch, before the void setup ()function. We can change the value of the variable anywhere in the code that is appropriate.

A *floating-point type* number can have a value from 3.4028235E+38 to as low as -3.4028235E+38. Each float has 6 or 7 digits and an exponent. Hopefully, you will not encounter a sketch where you need a bigger number.

***Try these experiments*** (think about what you expect to see before you do it and see if it comes out the way you expect):

1. Create an integer and make it equal to 3200, then print it to the serial port, over and over again.

2. Create a floating point variable equal to 25.991 and print it to the serial monitor.

3. Create a new floating-point variable equal to 22.4 and an integer, equal to 10. Then, assign the integer to equal the floating-point number. Print them both out.

## 5.6 New Features: Variable names

Variables are key elements in every sketch. The hardest part in using variables is thinking of *good names* for them. While we can use almost any name we want, I have certain guidelines that I use based on years of coding in many different languages.

The constraint on variable names is that they can't start with a number and can't have spaces. The only non-letter or number they can use is an underscore, "_".

For example, perfectly good variable names are:

- *a*
- *a5*
- *LED_pin*
- *b4*

In the olden days, with very limited memory in a computer, variable names were restricted to a single letter and a number. It was always a challenge trying to remember was the temperature sensor value variable name T3 or S2?

But these days, even in an Arduino, variable names can be 64 characters long or longer. So how do we design a good variable name?

As a *Best Design Practice*, the goal in selecting a variable name is to keep it short yet encode valuable information so it is self-documenting. Some of the information we could encode in the variable name are:

- *what type of variable: long, float, array, string...*
- *what element of the Arduino it refers: a pin, an input, out,...*
- *what the source or use of the number might be- as a specific sensor input, an LED value, a tone value...*
- *the units of the number stored in the variable*

Good variable names encode the most useful information so that you can remember what they mean a month later, or someone new to the code can decipher what the variable might contain.

If you use a consistent naming approach, your variables will be easy to decipher without having to check in the comment line, easy to remember, and easy to figure out if you don't want to find where it is declared. We want to establish good variable naming habits that are *self-documenting*.

Everyone has a different style when naming *variables*. This makes trying to figure out what is a variable and what type it is, just from the name, can be very confusing when reading someone else's sketch.

Here are two very good descriptions of variable naming conventions:

http://codebuild.blogspot.com/2012/02/15-best-practices-of-variable-method.html

https://dev.to/mohitrajput987/coding-best-practices-part-1-naming-conventions--class-designing-principles


Generally, I like to use variable names that describe the type of variable, in what general and specific context it is used and include the units of the number contained in the variable location.

To describe an integer, I like to start with the letter "i" or "n", like iCounter, or nptsAve_volts.

The rest of the name describes what the variable is used for and ends with the units. Here are some examples:

nCountCycles

iTimeStart_msec

iTimeStop_usec

sensorTemp1_volts

sensorTemp1_degC

iSensorTemp1_ADU

SensorTemp1_ADU

pinTemp1_hi

pinTemp1_lo


Any variable that refers to a pin name is unambiguously an integer, so starting with the word "pin" is ok.

The last part of any variable name is the units of the variable. To highlight the units, I set them off with an underscore, _, at the end of the variable name. If it is a time, the units might be _sec, or _msec, or _usec. If it is a voltage, it might be _volts or just _V or _mV.

In the special case of the levels from an *analog pin*, there are no units. The value is dimensionless. However, to help me keep track that the variable is a number from an analog pin, I use the units of *Analog to Digital Units* or _ADU. I always place the characters "_ADU" at the end of a variable name that will store the value read from an analog channel.

Even though a number in units of ADU, the number read from an ADC (Analog to Digital Converter), is an integer, sometimes we will make these float type numbers so we can do more accurate math, like when taking averages.

If the variable is used as a simple index number inside of a loop or in an array, which has meaning only as an index number, I would just use the letter i or i1 or i2 or i3.

> *There are no hard and fast rules. The Best Design Practice is to use variable names that help to describe what the variable refers to so anyone reading your code would have a clear idea. The second-best approach is to be consistent.*

It's important to develop good habits early on so that each sketch is an opportunity to practice and so that all of your projects become *self-documenting*.

You can never add too many comments to further clarify each variable.

Let's practice some of these variable principles.

***Try these experiments***:

1. What would be a good variable name for the pin driving an LED?

2. What is 20/3?

3. What is 20/3.0?

4. What would be a good variable name for the voltage of a sensor, stored as millivolts?

5. What is a good variable name to store the count of the number of times pin 12 has changed from high to low?

## 5.7 On-time, off-time, period and duty cycle

With our new skill at using variables, we can re-write the blink sketch with variables and make it a general pulse train synthesizer.

We will create a repeating pattern of on-times and off-times. We can describe this pattern in one of two ways:

- *in terms of an independent on-time and an off-time*
- *in terms of a repeat period and a fraction of the time it is on, which we call the duty cycle*

I like using the period and duty cycle to control the LED. This is a more intuitive set of figures of merit. To modulate the LED, and use the principles of the BLINK code, we will need to translate the *period* and *duty cycle* into the on and off times.

This is very simple. The on-time and off-time are:

onTime_sec = Period_sec x dutyCycle

offTime_sec = Period_sec x (1-dutyCycle)

The actual voltage to the LED and the pattern of on-and-off times is illustrated in Figure 5.3.

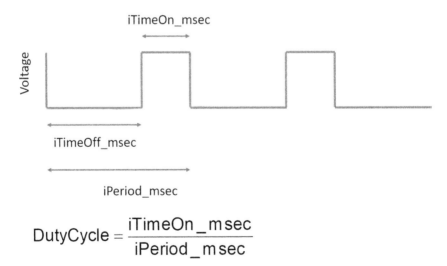

$$DutyCycle = \frac{iTimeOn\_msec}{iPeriod\_msec}$$

*Figure 5.3. The pattern of on and off times of the LED when it is modulated.*

The duty cycle controls the apparent brightness of the LED. A duty cycle of 90% means the LED is on for most of the period and it

## 5.7 On-time, off-time, period and duty cycle

will look bright to us. A duty cycle of only 10% means the LED is on for only a small fraction of the period and it will look dim to our eye.

We will modulate the LED with a period that is short enough so that our eye can't tell the LED is pulsing off and on.

To make the LED flash fast enough so we can't see it flashing, we want the period to be shorter than about 40 msec, or the flash frequency to be faster than 1/40 msec = 25 Hz.

This means we can adjust the on-time from 0 msec to 40 msec. We use delay() commands to adjust the on and off times.

A modified BLINK is all we need to drive the LED with a pulse train. We'll start with a period of 0.040 seconds (40 msec) and adjust the duty cycle from 0% to 100%.

***Try this experiment***: How would you write the code to set the period and duty cycle for the LED using variables? Use a period of 1 second and adjust the duty cycle. This way you can see the fraction of the time the LED is on.

Make the period shorter and shorter until it is shorter than the flicker rate and watch the LED transition from flashing to appearing on continuously.

*If you get stuck, watch this video and I will walk you through setting up this experiment.*

## 5.8 My sketch to modulate the LED with a period and duty cycle

```
// introducing variables
long iPeriod_msec = 500;
float dutyCycle_fraction = 0.5;

long iTimeOn_msec = iPeriod_msec * dutyCycle_fraction;
long iTimeOff_msec = iPeriod_msec - iTimeOn_msec;

void setup() {
  pinMode(13, OUTPUT);
}

void loop() {
  digitalWrite(13, HIGH);
  delay(iTimeOn_msec);
  digitalWrite(13, LOW);
  delay(iTimeOff_msec);
}
```

## 5.9 Summary of the commands introduced so far

| Command | Description |
| --- | --- |
| void setup(){ <br> } | This is a function that appears in EVERY sketch. Every command within the brackets will be executed just once |

## 5.9 Summary of the commands introduced so far

| | |
|---|---|
| `void loop(){` <br> `}` | This is a function that appears in EVERY sketch. Every command within the brackets will execute over and over again. |
| `pinMode(13, OUTPUT);` | This command tells the Arduino that we are going to use a specific digital pin as an OUTPUT, as distinct from an INPUT. We use the pin number to identify the pin we want to setup. |
| `digitalWrite(13, HIGH);` | This command controls the output of a digital pin and makes it either a HIGH (logic 1, or 5 V) or a LOW (a logic 0 or a 0 V). Once executed, the pin value will be set to this value. |
| `delay(1000);` | This command tells the Arduino to sit there, twiddling its thumbs, doing nothing for a duration of milliseconds as listed inside the (). In this example, the time interval to wait is 1000 msec = 1 sec. |
| `//` <br> or <br> `/*` <br> `*/` | Comments. Everything after these two forward slashes will be ignored by the Arduino sketch. We can use this add comments for our own benefit on any line. <br><br> For multiple lines add them between the /* and the */ lines. |

| | |
|---|---|
| `Serial.begin(9600);` | This command opens up the serial com link with a baud rated inside the (). In this example the baud rate is 9600 baud. The name `Serial` is an object, the serial link. The `.begin` is a verb, telling the object to get set up. |
| `Serial.print("Hello World");` | This command will print characters to the serial communications channel. Everything inside the () will be printed. To print a string, enclose the characters within quotes. After printing, the cursor will be left on the same line. |
| `Serial.println(" ");` | Does the same thing as the `Serial.print` command, but moves the cursor to the next line after printing the characters. This starts the next printed content on a new line. |
| Serial monitor | The terminal emulator that you can print to. This is the primary way of displaying information from the Arduino. |
| `int, long` | This will create a new variable that will store an integer. This is a whole number. It is usually placed before the setup() function |

## 5.9 Summary of the commands introduced so far

| | |
|---|---|
| `float` | This will create a new variable that will store a floating point number. This is a number that has a decimal point. It is usually placed before the setup() function |

# Chapter 6. Plotting Patterns in Numbers

The microcontroller at the heart of the Arduino is really a rather powerful computer. It is capable of a lot of calculation power. In this experiment, we will use the microcontroller as a computer to calculate patterns in numbers.

To look at these series of numbers, we will print them on the serial monitor. But I find it is difficult to look at patterns in a bunch of numbers. To see the patterns, we will introduce the serial plotter, a built-in graphing tool, to automatically plot the numbers as they are calculated.

## 6.1 What you need to know and what you will learn in this experiment

If you can run the Blink sketch and modulate the on-board LED with a duty cycle and period, you can write all the sketches in this chapter.

In addition to all the commands we've seen so far, we will introduce a few math functions to help us create patterns in numbers. We will continue introducing new ways of thinking of algorithms and how to turn them into code.

## 6.2 Printing columns of data

We can organize multiple numbers on one line by successive uses of the Serial.print() command. We use the Serial.print() command

## 6.2 Printing columns of data

to print the character that delimits or separates the numbers on one line.

The delimiter could a space, " ", a comma, ", ", or even a tab character. To print the tab character, we use the special command, Serial.print("\t").

Each number we print would have this delimiter between them until we get to the end of the line and the last number would be printed using Serial.println().

If we want to display three numbers on the same line, separated by commas, we could do something like,

```
Serial.print(5);Serial.print(", ");
Serial.print(10);Serial.print(", ");
Serial.print(15);
```

This would print the line:

5, 10, 15

Notice that I used two commands on one line. The commands are short, and they are very related. It's a simple way of organizing the code compactly and neatly.

This command is perfect if we want to output three columns of numbers that would be pasted into an excel spreadsheet, or even plot three different traces on the serial plotter, which will be introduced later in this chapter.

In the code above, successive print commands will keep printing on the same line, with the numbers right next to each other.

If we want to start printing on a new line, the last thing we print out should use the Serial.println() command. The "ln" at the end of the print command really stands for *linefeed*. It tells the printer command to, "*move the cursor to the next line, after the last item prints*".

In this example, to start the next set of three numbers on a new line, we would use:

```
Serial.print(5); Serial.print(", ");
Serial.print(10); Serial.print(", ");
Serial.println(15);
```

***Try this experiment***: use similar code as above and explore the difference between using the `Serial.print` command and the `Serial.println` command in the last line. Open the serial monitor to see the printed results.

## 6.3 Generating a series of numbers

Every sketch has a `void loop` (). We can take advantage of this loop to generate a continuous series of numbers defined by a few parameters or *figures of merit*.

We introduce a variable that will keep track of the index number, which will increment each time we loop. The way we do this is to initialize the index number before the loop and every time we go through the loop, we increment this value by 1 or some other interval.

I like using a variable name of "indexVal". It's rather boring but is simple and descriptive. It also indicates that it is a variable and not some special term called "index."

We take advantage of the structure of the built-in loop command to continuously change the value of a number each time through the loop.

***Try this experiment***: How would you write a sketch that creates an index variable, increments it each time the loop runs and then prints two columns of numbers, the index and the index squared, each number separated by a comma?

If you get stuck, look at my sketch in the next section.

This technique is a powerful method of automatically advancing a number incrementally using the built-in loop function.

If we can describe how a number will vary as a function of an index term, we can create its pattern and generate values for the pattern, forever.

*Watch this video and I will walk you through printing two columns of data.*

## 6.4 My sketch: incrementing an index and printing index, index$^2$

```
// practice printing number
long iVal_index;

void setup() {
  Serial.begin(2000000);
  iVal_index = 0;
}

void loop() {
  iVal_index++;
  Serial.print(iVal_index); Serial.print(", ");
  Serial.println(iVal_index * iVal_index);
  delay(200);
}
```

An example of the output of this sketch is shown in Figure 6.1.

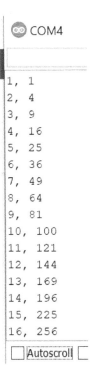

*Figure 6.1. The output in the serial monitor showing the index value and its square calculated by the microcontroller.*

## 6.5 An excel trick to place all the data into separate columns

We have just plotting two columns of numbers to the serial monitor. In this case, they are separated by a comma. We can literally copy the numbers from the serial monitor and paste them into an excel spreadsheet.

## 6.5 An excel trick to place all the data into separate columns

To make it easier to grab the numbers, especially the ones in a specific region of the printout, you should uncheck the "Autoscroll" box on the lower left of the screen. This will keep the numbers from moving up in the screen and make it easier to manually scroll through the list to find the specific ones to copy out.

In most spreadsheets, when you paste the data into the sheet, it all appears in one column. Usually, the paste operation does not separate the numbers into columns. However, in MS Excel, with one mouse click, we can convert the text into columns.

In MS Excel, this command is found in the Data menu, as the ***Text to Column*** command button.

The plot in Figure 6.2 is the graph that results from taking these columns of numbers, pasting them into excel and inserting a graph. It can be cleaned up to add axes labels and maybe changing the scaling or turn it into a line plot.

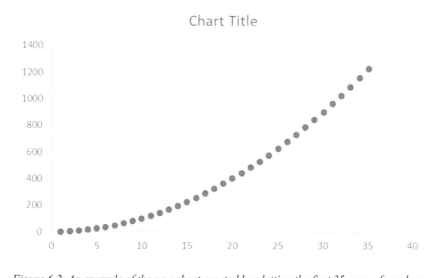

*Figure 6.2. An example of the raw chart created by plotting the first 35 rows of numbers copied from the serial monitor and pasted into excel.*

In this experiment, we just used the Arduino as a computation engine to calculate the numbers in a simple pattern and bring them into excel to use the power of the spreadsheet in plotting the result.

> *Watch this video and I will walk you through the process of creating this sketch, opening the serial monitor and pasting the data into excel.*

Now we have a simple way of generating example data to test out any plotting or data logging processes we want to explore.

## 6.6  New Code: the if...else command

To help control the flow of the code, we introduce a new command, the if statement. The syntax is

```
If (x == 1) {
  //do something
}
else {
  //do something
}
```

Inside the () is the statement (x==1). This is not a misprint. There really are two "=" signs.

When we used the "=" symbol before, it was always to assign a value to a variable, like iVal_add = 1.

But, in the if statement, we are not assigning a value to x, we are asking the question, is x = 1? This is a *conditional test*. The special symbol to do a *conditional* comparison is a double equal sign, ==. It is used as a new symbol, not some fancy double equal process.

When the sketch gets to this command, it will evaluate the statement inside the (). If it is true, the commands enclosed between the brackets, {}, will be executed. In this example, if x equals 1, the code in the brackets is executed.

If it is not true, the commands after the else statement will be executed. The else statement is optional. We can leave it out. If it is not there and the statement inside the () is false, the flow of the execution moves to the next command after the closing curly brackets, "}".

We can use this command to control the flow of commands, to add *structure* the code. When a value reaches a limit, we can execute some code that might reset, or change a number.

For example, suppose we want to create a ramp function. Here is a possible algorithm:

1. *We define iVal_increment as the increment to increase the ramp value by each time we loop. iVal_ increment could be +1 and the ramp value increases. An iVal_increment = -1 would make the values of the ramp term descend to smaller values.*

2. *Each time we loop, we increase the ramp value by an amount, iVal_increment.*

3. *When the ramp value reaches some maximum value, we change the sign of the increment value so the index ramps down.*

4. When the ramp value reaches a minimum value, we change the increment back to a positive value to ramp up.

***Try this experiment***: how would you implement this algorithm in a sketch and generate a continuously increasing and then decreasing ramp, in the shape of a triangle?

*Watch this video and I will walk you through the process of writing the sketch to create a triangle waveform*

## 6.7 My Sketch to create a ramp up and down or a triangle pattern of numbers

```
//triangle wave
long iMax_val = 100;
long iMin_val = 10;
long iTriangle_val = 50;
long increment_val = +1;

void setup() {
  // put your setup code here, to run once:
  Serial.begin(2000000);
}

void loop() {
  if (iTriangle_val == iMax_val) {
    increment_val = -1;
  }
  if (iTriangle_val == iMin_val) {
    increment_val = +1;
  }

  iTriangle_val = iTriangle_val + increment_val;
```

```
  Serial.println(iTriangle_val);
  delay(20);
}
```

There are two important features to notice in my code.

This ramp or triangle pattern has four figures of merit which I placed at the beginning of the sketch. This way they are located in a convenient place in the sketch making them easy to find and to change:

```
long iMax_val = 100;
long iMin_val = 10;
long iTriangle_val = 50;
long increment_val = +1;
```

The first two terms determine the range of values we will see in the calculated numbers.

The third and four terms are initial values. These define how we start the ramp. Using a value of iVal_increment that is positive means that we start the ramp with values increasing. And, the first point we start at is iVal_ramp = 1 in this case.

Right now, we are starting near the beginning of the ramp pattern. If we make its value closer to the max value, we would be starting near the peak of the ramp.

The second feature in my code is that I am using the baud rate of 2000000, which I will do in all cases going forward. If it is reliable, why not use the highest data transfer rate I can?

However, these are features I chose to include in my sketch. When there are multiple right ways of writing the code you get to express your own creativity. Maybe you did it differently?

This is why it's fun to see how others solved the same problem you did.

*Sometimes you will find you came up with a more clever or simpler or faster way of implementing the solution that another person. Sometimes you will learn a new trick from another's sketch.*

When we use the min and max values of 0 and 10 and run this code, we will see the numbers printed to the serial monitor increase linearly from 0 to 10 and then ramp down to 0 and then ramp back up, repeating forever.

Here is a small snippet of output on the serial monitor, shown in Figure 6.3.

Figure 6.3. Example of the printout for the ramp generator.

While we see a ramp pattern in the numbers on the screen, they are still just a bunch of numbers. It would be cool to see these numbers plotted on a graph.

## 6.8 New Feature: Plotting the data in the IDE with the Serial Plotter

In the last example, we generated one column of data in the serial monitor. It was just a series of numbers in a list. It would be nice to see the pattern these numbers created on a graph.

We can use a built-in feature in the Arduino IDE to automatically plot any series of numbers, the *Serial Plotter*.

When the Arduino sends a sequence of numbers on the serial com port, the Serial Plotter will plot the numbers as the data comes in.

Very conveniently, the Arduino IDE allows plotting not just of one trace but as many as six different sets of data all at the same time. Each one will be plotted in a different color, BUT, on the same vertical and horizontal scale.

To plot multiple traces at the same time on the same plot, we just print each number separated with a comma, ",". If we want to plot the variables a1, b1, c1, d1 and e1, we could use these lines in the loop:

```
Serial.print(a1); Serial.print(", ");
Serial.print(b1); Serial.print(", ");
Serial.print(c1); Serial.print(", ");
Serial.print(d1); Serial.print(", ");
Serial.println(e1);
```

Each line prints the variable followed by a comma and space, and then the next variable. The last line moves the cursor to the next line on the serial monitor, or the next set of points to be plotted.

There are a few limitations to the serial plotter.

All the traces will be plotted on the same scale. If one trace had values that vary from 1 to 2 and another uses values that vary from -100 to +100, the scale will be adjusted for the -100 to +100 trace. This means the small value trace will look mostly like a straight line.

The second limitation is that each of the values printed on the same line, without a line feed between them will be plotted with the same horizontal value, one of the 500 locations on the time axis. This is a good feature in that it automatically synchronizes when each point in each trace is plotted.

On the Serial Plotter, a total of upto 500 points are plotted across the screen at any one time. After the 501st point is sent to the com port, the plotted data points shift to the left to make room for the next one. As more numbers come over the comm link to the Serial Plotter, each new point is added to the right and the whole curve shifts to the left.

We access the Serial Plotter from the Tools menu of the IDE, as seen in Figure 6.4.

## 6.8 New Feature: Plotting the data in the IDE with the Serial Plotter

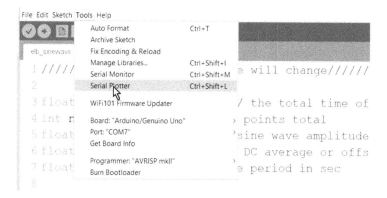

*Figure 6.4 Access the Serial Plotter from the tools menu. Be sure the Serial Monitor is closed.*

We can use the previous sketch to generate a pattern of repeating numbers plotted on the Serial Plotter. I ran the previous sketch that generated a ramp or triangle pattern, using parameters of:

```
int iVal_max = 30;
int iVal_min = 0;
int iVal_increment = +1;
int iVal_ramp = 1;
int iDelay_msec = 10;
```

I got the initial pattern shown in Figure 6.5.

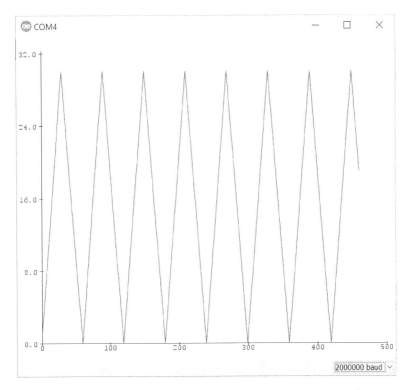

*Figure 6.5. Initial pattern of numbers displayed on the serial plotter as they come across the serial comm link in real time.*

The serial plotter will plot the first 500 points on the screen, and then scroll to the left as more data points are plotted. This will continue forever.

***Try this experiment***: Using the ramp triangle function as the source of data, think about how each of the five figures of merit will change a feature in the displayed pattern. Try varying them and see if their impact is what you expect.

Adjust the delay time to control the rate at which the data is plotted. If a 1 msec delay is used, it will take 500 x 1 msec = 500 msec = 0.5 seconds to plot a screen's worth of data. Using a delay of 4 msec will take 500 x 4 msec = 2,000 msec = 2 seconds to plot a screen's worth of data.

***Try this experiment***: How would you expect the plot to change if you adjust the max value? Try some extreme values to explore the properties of the serial plotter and our method of generating a ramp function.

***Try this experiment***: Supposed you wanted to plot one screen's worth of data as fast as you can, and then pause for 5 seconds to view the plot. How would you write this code using if statements? *Hint*: consider adding a counter variable to keep track of how many points have plotted.

*One caution*. You can only have the serial monitor OR the serial plotter open at one time. If the serial monitor was open, you have to close it before you can open the serial plotter, and vis versa. Otherwise you will get an error message.

## 6.9 New Feature: Auto format is your friend

If you were to type the sketch below, It's Alive, into a blank sketch, or just copy and paste it into a blank sketch, it will run. But just won't look very neat:

```
//it's alive
void setup() {
pinMode(13, OUTPUT);
}

void loop() {
digitalWrite(13, HIGH);
delay(100);
digitalWrite(13, LOW);
delay(100);
digitalWrite(13, HIGH);
delay(100);
digitalWrite(13, LOW);
```

```
delay(500);
}
```

This is the It's Alive sketch we introduced earlier. Figure 6.6 shows a copy of the sketch as it looks directly pasted into a blank sketch.

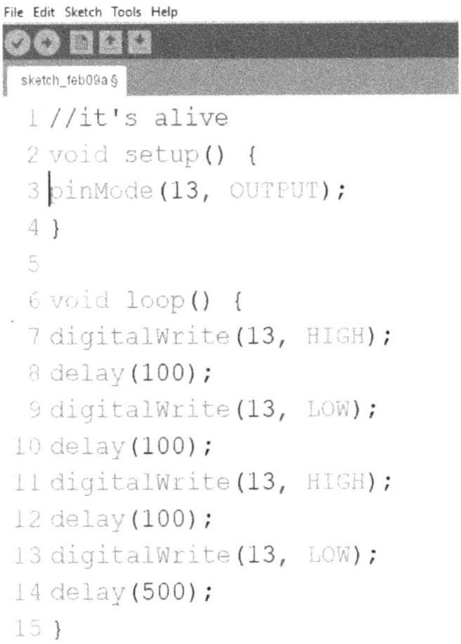

```
File Edit Sketch Tools Help

sketch_feb09a §
 1 //it's alive
 2 void setup() {
 3 pinMode(13, OUTPUT);
 4 }
 5
 6 void loop() {
 7 digitalWrite(13, HIGH);
 8 delay(100);
 9 digitalWrite(13, LOW);
10 delay(100);
11 digitalWrite(13, HIGH);
12 delay(100);
13 digitalWrite(13, LOW);
14 delay(500);
15 }
```

*Figure 6.6. When you type the It's Alive sketch into a blank screen, it will run perfectly, but does not look very neat.*

We can make this look a little neater by indenting each line that is within a function or a loop statement. Rather than do this manually, we can let the IDE (Integrate Development Environment) do it for us.

Under the *Tools* menu is the *Auto format* command. This is shown in Figure 6.7.

## 6.9 New Feature: Auto format is your friend

*Figure 6.7 The autoformat command is under tools. This will make you code much neater and easier to read.*

After applying this auto format tool, the code is nicely indented, shown in Figure 6.8.

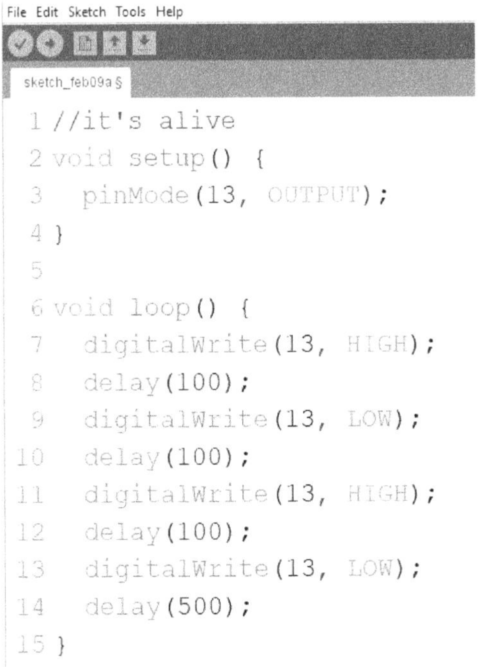

```
1 //it's alive
2 void setup() {
3   pinMode(13, OUTPUT);
4 }
5
6 void loop() {
7   digitalWrite(13, HIGH);
8   delay(100);
9   digitalWrite(13, LOW);
10  delay(100);
11  digitalWrite(13, HIGH);
12  delay(100);
13  digitalWrite(13, LOW);
14  delay(500);
15 }
```

*Figure 6.8 After autoformat, the code is automatically indented and easier to read.*

## 6.10 Advanced: Controlling the number of decimal places in the Serial.print() command

*Watch this video and I will walk you through*

One of the most common ways of outputting the value of a variable is with the Serial.print(variable_name) command. When we print an integer, we see all of its digits. But, when we print a

## 6.10 Advanced: Controlling the number of decimal places in the Serial.print() command

floating-point number, we only get two of the digits printed by default.

This is illustrated in the following small sketch:

```
void setup() {
  Serial.begin(2000000);
}
void loop() {
  Serial.print(25); Serial.print(",   ");
  Serial.println(25.123456);
}
```

The number that is printed is 25.12. This isn't good enough if we are printing a voltage, for example. We want to see the third digit, the mV value.

Here is the trick to tell the Serial.println() command to print more digits for a floating-point number.

Inside the () of the print command, we give the variable name or value, 25.12345, and the number of digits we want to print separated with a comma, ",". Here is an example:

```
void setup() {
  Serial.begin(2000000);
}
void loop() {
  Serial.print(25); Serial.print(",   ");
  Serial.println(25.123456,3);
}
```

In this sketch, I will print out three digits on the screen as a result of the print command.

***Try this experiment***: Run the sketch above and see the number printed, then change the 3 to a 4 in the Serial.println() command.

You should see 25.1235 as the number printed out. It is 4-digits, but the last one displayed is a rounded-up value of the remaining digits.

We are now armed with the ability to print data to the serial monitor at the highest data rate and manipulate the format of the information on the screen.

## 6.11 Summary of the commands introduced so far

| Command | Description |
| --- | --- |
| `void setup (){`<br>`}` | This is a function that appears in EVERY sketch. Every command within the brackets will be executed just once |
| `void loop (){`<br>`}` | This is a function that appears in EVERY sketch. Every command within the brackets will execute over and over again. |
| `pinMode(13, OUTPUT);` | This command tells the Arduino that we are going to use a specific digital pin as an OUTPUT, as distinct from an INPUT. We use the pin number to identify the pin we want to setup. |

## 6.11 Summary of the commands introduced so far

| | |
|---|---|
| `digitalWrite(13, HIGH);` | This command controls the output of a digital pin and makes it either a `HIGH` (logic 1, or 5 V) or a `LOW` (a logic 0 or a 0 V). Once executed, the pin value will be set to this value. |
| `delay(1000);` | This command tells the Arduino to sit there, twiddling its thumbs, doing nothing for a duration of milliseconds as listed inside the (). In this example, the time interval to wait is 1000 msec = 1 sec. |
| `//` or `/*` `*/` | Comments. Everything after these two forward slashes will be ignored by the Arduino sketch. We can use this add comments for our own benefit on any line.<br><br>For multiple lines add them between the /* and the */ lines. |
| `Serial.begin(9600);` | This command opens up the serial com link with a baud rated inside the (). In this example the baud rate is 9600 baud. The name `Serial` is an object, the serial link. The `.begin` is a verb, telling the object to get set up. |

| `Serial.print("Hello World");` | This command will print characters to the serial communications channel. Everything inside the () will be printed. To print a string, enclose the characters within quotes. After printing, the cursor will be left on the same line. |
|---|---|
| `Serial.println(" ");` | Does the same thing as the `Serial.print` command, but moves the cursor to the next line after printing the characters. This starts the next printed content on a new line. |
| Serial monitor | The terminal emulator that you can print to. This is the primary way of displaying information from the Arduino. |
| `int, long` | This will create a new variable that will store an integer. This is a whole number. It is usually placed before the setup() function |
| `float` | This will create a new variable that will store a floating-point number. This is a number that has a decimal point. It is usually placed before the setup() function |
| `+, -, *, / =` | Simple algebra |
| `==` | A conditional test: are the two terms on either side of the == the same. Not a misprint, double equal signs. |

## 6.11 Summary of the commands introduced so far

| | |
|---|---|
| `if (iVal == 1) {`<br>`}` | If conditional. If the statement inside the () is true, then the commands between the brackets are executed. |
| Tools/Autoformat | Will add indents and clean up the look of your code |
| Serial Plotter | A built-in graphing features which plots the numbers printed to serial com port. |

# Chapter 7. How Random are Random Numbers?

In this experiment, we will explore some of the properties of random numbers using a built-in random number generator.

## 7.1 What you need to know and what you will learn in this experiment

We continue the development of new algorithms and using the Arduino as a computation engine.

We will leverage the skills you learned in the previous chapter calculating patterns in numbers and plotting them. Being able to perform calculations and plot the numbers is important in this chapter.

## 7.2 New command: the for loop

One of the most important coding features we use is a *loop*. This is sometimes referred to as a *control structure*, as it controls the *flow* of the code.

In the last example of generating the triangle or ramp pattern, we used the built-in loop that is part of *every* sketch. Now we will look at loops we add to a sketch.

A loop is a special function that will execute the same group of commands over and over again, controlled by a condition to get out of the loop. Once we enter the loop, the only way we get out of it is if a condition we specify is met.

If we never meet the condition, we could stay in a loop forever. Sometimes this is intentional, sometimes it is by accident.

There was a Star Trek Next Generation episode in season five, Cause and Effect, where the Enterprise got caught in a time loop and it was up to Data to get them out of the loop.

If you don't have a "Data" in your sketch, you could get trapped in the loop and never move to the rest of your sketch. These "infinite" loops are sometime difficult to debug. Be on the lookout for them when your sketch does not go as planned.

There are three types of loops. They vary by how we specify stopping the loop:

- ***for loop***: *end the loop after looping n times*
- ***do...while***: *end the loop based on a truth statement we execute at the end of the loop*
- ***while***: *end the loop based on a truth statement we execute at the beginning of the loop*

In this experiment, we will create a fixed number of random numbers and do things to them. We could use any of these three types of loops to create the fixed number of points, but the for loop will be the easiest to use.

The *for loop* has the following structure:

```
for (index = 1; index <= npts; index = index + 1) {
  //execute this code if index<=npts
}   //after the for loop completes, code moves here
```

The ***first term***, inside the () of the for command, index = 1, sets up the beginning of the loop. Here, the term, index, is a variable name which I declared as an integer somewhere else in the program.

The variable labeled *index*, is the counter. Stored in this variable is the value of the number of times we have gone through and executed the loop.

If we are only using this variable in this for loop and nowhere else, we can actually create it right in the for loop. You will often see sketches with the first line as "int index = 1;".

This creates the variable as an int and initializes it to 1 all at the same time.

The **second term**, index <= npts, is the condition tested in each iteration of the loop. As long as this condition is *true*, the loop continues to execute. As soon as it is tested and found to be *false*, the execution flow moves to after the ending curly bracket.

The **third term**, index = index + 1, is how we increment the index counter. Inside the loop, we will use the index term in our calculation so that each time we run through the loop, we will use the new index value to generate a new number.

This operation of incrementing an index term or counter by 1 every time the loop executes is so common, there is a shortcut created just for this.

Instead of writing index = index + 1, we just write index++. The "++" means increment 1 to this variable. This is how the programing language C++ gets is name. It is the next iteration of C.

***Try this experiment***. Remember in the last chapter, we saw that we can plot only 500 points on the serial plotter at any time. How would you write the code to plot the first 500 points in the series of numbers from 1 to 500, wait 5 second to admire your handiwork and replot from the beginning?

If you get stuck, take a look at my sketch below:

```
void setup() {
  Serial.begin(2000000);
}

void loop() {
  for (int i = 1; i <= 500; i++) {
    Serial.println(i);
  }
  delay(5000);
}
```

## 7.3 Random numbers

In this experiment, we are going to build a *random number generator*, and explore some of its properties. This is a useful tool when we want to add a little unexpectedness to a series of numbers or just see a constantly changing pattern.

The random() function will generate "pseudo random" numbers. But they are random enough for our purposes.

The syntax of the function is

`i1=random(min, max);`

Every time this code is executed the value of i1 would be a different random number between the minimum value and the (max value – 1). It will return a long integer.

If we want a random number between 0 and 1000, we would use random(0, 1001). The default value for the starting number is 0, so this can be shortened to random(1001).

***Try this experiment***: How would print out 500 random numbers between 0 and 100. After you display them on the serial monitor, try plotting them. Think about what you expect to see before you plot them. Is what you see, what you expected?

Hint: just one line in the for loop needs to be changed. The random function can be placed directly into the Serial.println () command.

## 7.4 Random walking

The random () function will return a "*random*" integer with a value between two numbers set as the limits.

Every time this function is called, it will return another random number. But how random is it really?

There many tests for the randomness of a sequence of numbers. The best list of tests is from the NIST. They will be the first to say that testing for randomness is really hard. In fact, there is no definitive, absolute test for randomness. You would literally have to check an infinite number of samples to know if it were really random.

However, there are a few simple, practical tests that at least give an indication. My favorite test is the *cumulative sum*, as compared to a *random walk* prediction.

A random walk is a path taken, controlled by the random flip of a coin. It is sometimes called a "drunkards walk" for reasons that will soon be obvious.

To create a random walk, we walk on a straight line that is oriented vertically. For each step, we decide whether to move up or down on the line, based on a coin flip. For example, a head moves us up one step, a tail moves us down one step.

After each coin flip trial, we note our distance from the starting place and plot this. The horizontal axis on this plot is the step number. Our position on the line after each step is plotted on the vertical scale. This is a plot of a random walk, such as shown in Figure 7.1.

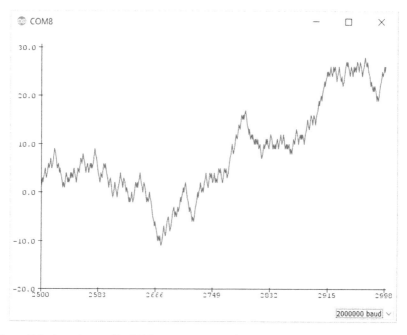

*Figure 7.1. A random walk of 500 steps starting at 0 on the vertical axis. At each step whether we move up or down is decided by the flip of a coin.*

To calculate the random walk plot we call a heads a +1 and a tails a -1. As we flip the coin, we keep a running sum. The running sum is the net position from the origin. If the flipping process is really random, then on average, we should see just as many heads as tails and the running sum should average to 0.

Without any experience in statistics, we might expect we would see our position fluctuating about the 0 value as we increase the number of coin flips. We should see an average value of 0 ± some noise.

But this is not the case.

## 7.5 Estimating deviation likelihood from 0 in a random walk

In practice, the net position from the start in a random walk will never be precisely 0. The net distance will fluctuate. How much fluctuation is expected as *normal*? A little bit of statistical analysis says that on average, we expect the deviation from 0 to scale roughly with the square root of the number of flips.

If we flip the coin 100 times, the average value should be 0, but we would not be surprised to see deviations by as much as ±10 from 0. As we flip more coins the expected deviation of the running sum from 0 would grow, but with the square root of the number of flips.

You might think, "but wait, if this is true, wouldn't the expected deviation from 0 grow as we flip more coins? Doesn't the actual average value get farther from the expected 0?" Yes, but the fraction of all the flips that might deviate from 0 will decrease. The relative fraction of the deviation is:

$$\text{deviation fraction} = \frac{\sqrt{n}}{n} = \frac{1}{\sqrt{n}}$$

As we flip more coins, we should see the fraction of how far away we are from 0 will decrease as the square root of the number of flips. This is something simple we can test.

The idea is to generate a series of random coin flips, each time getting a +1 or -1. We calculate the running sum and display it on a plot. We also display the expected deviation, which would be $\sqrt{n}$, and just as likely to be (+) as (-).

This sort of experiment is called a *random walk* experiment. The cumulative sum is random walking from 0 each time we flip the coin. It could move equally likely in the positive direction or the negative direction.

If there were a bias and a head or +1 was slightly more likely than a tail or -1, then we would see the cumulative sum always veering toward an increasing positive value.

## 7.6 An electronic coin flipper

There is an important property of the random number generator in the Arduino IDE. Every time you start the sketch, the random function will generate identical random numbers. The sequence will start over again each time you start the sketch.

You can test this out by running the random function and just printing out the series of numbers in repeated runs.

Remember, all these computations are done in the microcontroller itself. We are literally using the microcontroller as a computer to calculate these numbers, do some algebra with them and send them over the serial com link to have the computer just plot them up.

***Try this experiment***: How would you write the code to just print out a list of 10 random numbers from 0 to 9? Remember the syntax for the random function is:

```
random(0, 10);
```

Try this yourself before you look at my sketch.

> *Watch this video and I will walk you through writing the sketch to generate a list of 10 random numbers.*

The entire sketch to print 10 random numbers is just a few lines:

```
void setup() {
  Serial.begin(2000000);
}
void loop() {
  for (int i = 1; i <= 10; i++) {
    Serial.println(random(0, 10));
  }
  delay(10000);
}
```

When we execute this sketch three different times, we get exactly the same set of ten numbers each time we run the code:

| 7 | 7 | 7 |
| 9 | 9 | 9 |
| 3 | 3 | 3 |
| 8 | 8 | 8 |
| 0 | 0 | 0 |
| 2 | 2 | 2 |
| 4 | 4 | 4 |
| 8 | 8 | 8 |
| 3 | 3 | 3 |
| 9 | 9 | 9 |

While the numbers in each trial may be random, they are definitely not random from trial to trial.

If you want a completely random sequence that is different each time you start the sketch, you need to start the random sequence with a different *seed number*. This is the number that starts the random number sequence and makes it a unique sequence each time the random() function is called.

The command is

randomSeed();

The number inside the randomSeed() function will start the random sequence. Every time this number, which is in the form of a *long integer* number, changes, the random sequences generated by random () will be different.

We place the randomSeed() at the beginning of the sketch. Each time the sketch starts, we want this number to be different. This way the random sequence is different each time the sketch starts.

This suggests we want the number inside the function to itself be random or varying each time the sketch is started. I like using the time from the start of the sketch, in microseconds, plus whatever noise is on the analog pin. The command would be:

```
randomSeed(analogRead(A5)+micros());
```

If nothing is connected to the analog pin, there will always be some 60 Hz pick up and static field pick up and some other noise measured. In the next chapter we explore these sorts of measurements.

The sum of these two numbers will be pretty random, so the random seed number will be random and the random number sequence will be unique each time the sketch is run.

## 7.7 Calculating a random walk

Now we are armed with the tools we need to calculate a random walk. The algorithm we will follow is:

1. Generate two random numbers.

2. Turn them into a value of +1 or -1.

3. Calculate the sum for the first 500 trials, plotting the cumulative sum as you go.

4. Also plot the + sqrt(n) and - sqrt(n) to compare the cumulative sum with the expected ranges, where n is the number of the trial.

5. For extra credit, make each start of the sketch randomized.

When you are ready, take a look at the sketch I created.

*If you get stuck, watch this video of how I created the sketch.*

## 7.8 My Sketch: Plotting a random walk with limits

Here is the sketch I wrote to create 500 random numbers, calculate their cumulative sum and their deviation likelihood.

```
long iSum = 0;
/////////////////////////////
void setup() {
  Serial.begin(2000000);
  randomSeed(micros() + analogRead(A5));
```

```
}
void loop() {
  iSum = 0;
  for (int i = 1; i <= 500; i++) {
    iSum = iSum + (2 * random(0, 2) - 1);
    Serial.print(sqrt(i)); Serial.print(", ");
    Serial.print(-1 * sqrt(i)); Serial.print(", ");
    Serial.println(iSum);
  }
  delay(2000);
}
```

This sketch will generate random walk trials of 500 flips and plot their cumulative sum and their deviation likelihood on the serial plotter. They will stay on the screen for 2 seconds so you can take a look at how large the random walk is compared with the likelihood deviation estimate of $\sqrt{n}$.

An example of a typical run showing the random walk compared to the estimates is shown in Figure 7.2. The random walk excursions are mostly within the limit lines.

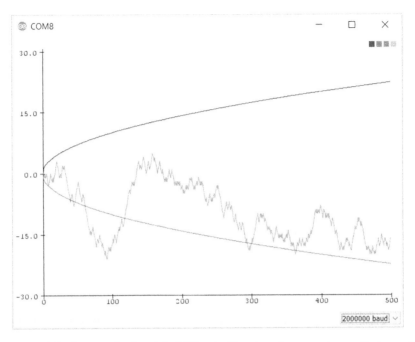

*Figure 7.2. An example of a trial of 500-coin flips plotted as a random walk along with the limit lines based on the square root of the number of flips.*

If you find the value is consistently higher than $+\sqrt{n}$ or lower than $-\sqrt{n}$, then there is reason to believe the random number generator is not random or there is some non-random influence.

***Try this experiment.*** Observe many sets of 500 trials. How often does the random walk pattern stay within the $\sqrt{n}$ limits or go outside the bounds?

1. How often would you say the random walk ends up positive, as compared with negative values?

2. How often does the random walk go outside the range expected for a random walk?

3. How well does the square root of n describe the expected range for the random walk?

## 7.9 A better metric for the worst-case likely deviation

What I find remarkable is that as we grow the number of flips, the deviation from 0 really does seem to scale with the square root of the number of flips. But, there are always deviations and sometimes the deviation is beyond just $\sqrt{n}$.

To judge how much beyond $\pm\sqrt{n}$ we random walk, it is sometimes useful to add another set of lines, which are $\pm 2 \times \sqrt{n}$.

***Try this experiment***: Add four scaling lines, one set at $\pm\sqrt{n}$ and one set at $\pm 2 \times \sqrt{n}$.

Here is my sketch:

```
long iSum = 0;
///////////////////////////
void setup() {
  Serial.begin(2000000);
  randomSeed(micros() + analogRead(A5));
}
void loop() {
  iSum = 0;
  for (int i = 1; i <= 500; i++) {
     iSum = iSum + (2 * random(0, 2) - 1);
    Serial.print(sqrt(i)); Serial.print(", ");
    Serial.print(-1 * sqrt(i)); Serial.print(", ");
    Serial.print(2 * sqrt(i)); Serial.print(", ");
    Serial.print(-2 * sqrt(i)); Serial.print(", ");
    Serial.println(iSum);
  }
  delay(2000);
}
```

Now when I look at the plots, I see the random walk values are almost always within the larger limit-lines. The probability of exceeding the $\pm 2 \times \sqrt{n}$ limits is very low. An example is shown in Figure 7.3.

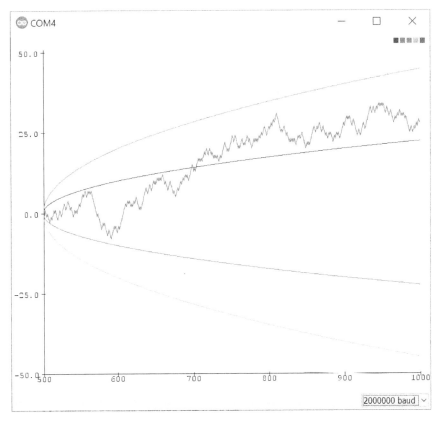

*Figure 7.3. An example of a random walk with the two limit lines corresponding to 1 and 2 x sqrt(n).*

## 7.10 What if we keep increasing the number of coin flips?

In the last sketch, we plotted the results of trials with 500-coin flips, because the serial plotter can plot only 500 points at a time.

## 7.10 What if we keep increasing the number of coin flips?

We will increase the trial size in multiples of 500 and plot the running sum so we see the cumulative running sum, sampled 500 times through the entire run. This is the maximum number of points that can be plotted on the serial plotter.

***Try this experiment***. Given the limitation of the serial plotter to only display 500 points at a time, how would you structure your code and write the sketch to calculate the cumulative sum of n points at a time, and then repeat this 500 times?

There are many right ways of performing this experiment. The way I did this is to introduce a new variable, nMultiple, as the number of multiples of 500 for the number of coin flips or trials.

The limit on how large to make nMultiple will be based on how long we want to wait to calculate and display the random walk. Afterall, all of these calculations are actually being done on that tiny microcontroller on the Arduino board.

We are using it a computer to perform all the calculations. As tiny as it is, it is about as powerful as a 1985 PC. Many times more powerful than the computer that landed astronauts on the Moon in July, 1969.

***Try this experiment***: Modify the code to increase the number of flips by nMultiple. In addition to the estimated excursions of $\pm\sqrt{n}$ add the estimates of $\pm 2 \times \sqrt{n}$.

Think about how you would implement this. If you get stuck take a look at my sketch.

## 7.11 My sketch: random walking with multiples of 500 coin flips

```
long iSum = 0;
int nMultiple = 10;
long iCount;
/////////////////////////////
void setup() {
  Serial.begin(2000000);
  randomSeed(micros() + analogRead(A5));
}
void loop() {
  iSum = 0;
  iCount = 0;
  for (int i = 1; i <= 500; i++) {
    for (int i2 = 1; i2 <= nMultiple; i2++) {
      iSum = iSum + (2 * random(0, 2) - 1);
      iCount++;
    }
    Serial.print(sqrt(iCount)); Serial.print(", ");
    Serial.print(-1 * sqrt(iCount)); Serial.print(", ");
    Serial.print(2 * sqrt(iCount)); Serial.print(", ");
    Serial.print(-2 * sqrt(iCount)); Serial.print(", ");
    Serial.println(iSum);
  }
  delay(2000);
}
```

To keep track of each flip to know what the value of $\sqrt{n}$, I added another variable, iCount, to keep track of the total number of flips.

Try different values for nMultiples to see how large we can go without taking too much time.

When I ran this experiment, I could use an nMultiple value of 50 and not take too long to calculate. This resulted in a total of 50 x 500 = 25,000 coin flips per run, plotted on the serial plotter.

Remember, once we upload all the code to the Arduino, all the calculations are being done on the tiny microcontroller chip.

I ran dozens of experiments and consistently found the deviation usually within $\pm\sqrt{n}$ and always within $\pm\, 2 \text{ x }\sqrt{n}$. An example of one run of 25,000 coin flips, using nMultiple = 50, is shown in Figure 7.4.

*Figure 7.4. Example of the random walk for 25,000 coin flips and the two limit lines for the expected deviation.*

**Try this experiment**: Does the random walk pattern look different if you use an nMultiple of 1 or 50? Does it really matter how many flips we measure? Will the randomness look the same?

## 7.12 Relative distance traveled

Am I the only one that finds it rather bizarre that the more flips of the coin, the farther away we move from 0? Afterall, the average value should be 0, so with more coin flips, shouldn't we get closer and closer to 0 in our calculation?

What makes our random walk so counter-intuitive is that we are plotting the distance we travel in the random walk as we increase the number of coin flips. If we just walked along the line linearly, each step in the same direction, we would progress 500 steps down the line in 500 flips of the coin.

As we flip more coins our linear distance would move down the line farther and farther. The deviation in the random walk would also increase, but not as fast. The deviation in the random walk would increase slower than the distance we would travel if our progress was linear and not random. The relative distance of our random walk, compared to the linear distance, would get smaller and approach a value of zero.

We can see this by plotting not the deviation likelihood, but the fraction of how large the random walk is compared to the linear walk, which is the index value, i.

In my sketch, I paid attention to four subtle features.

*First*, I wanted to look at a lot of flips so I started with the sketch from the last example, that allowed me to do 25,000 flips, plotted on our scale of 500 points horizontally.

*Second*, when I took the average, I wanted to use floating point math, not integer math. I converted the integer number of counts, iCount, into a floating point by multiplying it by 1.0.

*Third*, the smallest scale the serial plotter can plot is a range of 6 units. We will be plotting values much less than 1, so they will not appear very large on the screen. To expand them out a little, I scaled the fraction by 100 to use more of the screen. This way, the fractional value is really being plotted as a percentage.

*Fourth*, I added a plot of the 0 level to make it easier to compare the relative random walk to the 0 level.

## 7.12 Relative distance traveled

Here is my final sketch:

```
long iSum = 0;
int nMultiple = 50;
long iCount;
/////////////////////////
void setup() {
  Serial.begin(2000000);
  randomSeed(micros() + analogRead(A5));
}
void loop() {
  iSum = 0;
  iCount = 0;
  for (int i = 1; i <= 500; i++) {
    for (int i2 = 1; i2 <= nMultiple; i2++) {
      iSum = iSum + (2 * random(0, 2) - 1);
      iCount++;
    }
    Serial.print(iSum/(iCount*1.0)*100);
    Serial.print(", ");
    Serial.println(0);
  }
  delay(2000);
}
```

An example of the plot of this behavior is shown in Figure 7.5.

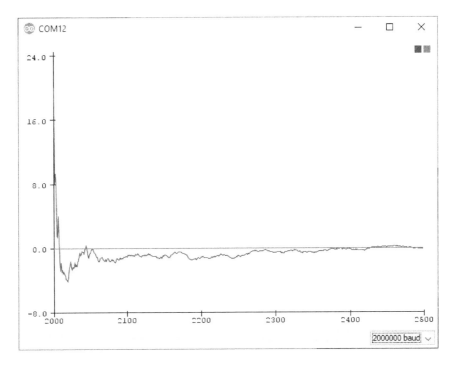

*Figure 7.5. The relative random walk compared to the linear walk does converge to 0 as we walk farther.*

After 25,000 flips, the expected deviation of the random walk is $\sqrt{25,000} = 158$. The relative percentage of this expected deviation from the linear value of 25,000 is 158/25,000 = 0.6%.

I would not be surprised to see final percentage deviations from 0 on the order of 0.6%. Completing more than a dozen trials, the final relative value was always less than about 1%, exactly as I expected.

We see the relative distance we travel in the random walk, as a fraction of a linear walk is converging to 0 with more steps, as we expect. The more we walk, the farther we random walk away from 0 on an absolute size, but the closer the relative distance is to zero.

## 7.13 Can you win at Roulette using this simple method?

We know that for any single coin flip trial, the probability of getting a head or a tail is exactly the same, 50%. On average, there will be the same number of heads and tails.

But, the probability of getting four heads in a row is 0.5 x 0.5 x 0.5 x 0.5 = 6%. It would seem that if we get three heads in a row, the probability of the next flip to be a head is only 6%. Does this mean that the probability of getting a tail on the next flip is 94%? This sounds like a sure bet.

We can modify our simple coin flip code to test this idea.

***Try this experiment***: How would you write the code to implement this experiment: Build a coin flipper and monitor the heads and tails. When there are three consecutive heads, how often is the fourth flip a head or a tail?

The algorithm might be:

1. *conduct multiple coin flips*
2. *wait until there are three consecutive heads*
3. *record the fourth flip*
4. *plot the random walk of the running sum of the fourth value after three consecutive heads*

If you get stuck, check out my sketch below.

## 7.14 My sketch to test out the sure bet approach to winning at roulette

Here is my sketch:

```
long iSum = 0;
int nMultiple = 10;
long iCount;
///////////////////////////
void setup() {
  Serial.begin(2000000);
  randomSeed(micros() + analogRead(A5));
}
void loop() {
  iSum = 0;
  iCount = 0;
  while (iCount < 500) {
    if (random(0, 2) == 1 and random(0, 2) == 1 and random(0, 2) == 1) {
      iSum = iSum + (2 * random(0, 2) - 1);
      iCount++;
      Serial.print(sqrt(iCount)); Serial.print(", ");
      Serial.print(-1 * sqrt(iCount)); Serial.print(", ");
      Serial.print(2 * sqrt(iCount)); Serial.print(", ");
      Serial.print(-2 * sqrt(iCount)); Serial.print(", ");
      Serial.println(iSum);
    }
  }
  delay(2000);
}
```

Note that I switched from a *for loop* to a *while loop*. The *for loop* runs for a specific number of iterations. In this case, though I wanted to record 500 final flips, ultimately, I would need a second criterion to decided when to record the next electronic flip. I didn't know how many flips it would take to get to three consecutive heads.

The great thing about these sorts of sketches is there are multiple ways of getting the same result. This means you can really apply your creativity in finding a solution.

We can *experimentally* test our idea of how to win at Roulette using this sketch. Here is the result of one run, shown in Figure 7.6.

*7.14 My sketch to test out the sure bet approach to winning at roulette*

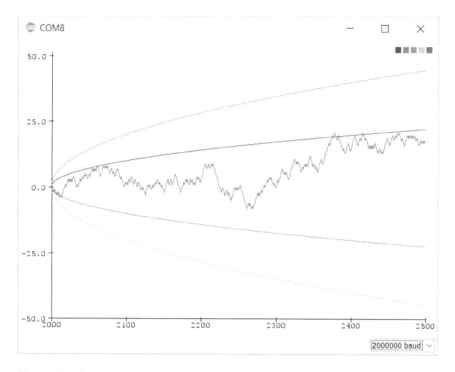

*Figure 7.6. The cumulative sum ONLY recording a flip if there are 3 consecutive heads first.*

Our idea doesn't seem to work. How can this be? The chance of having four heads in a row is only ½ x ½ x ½ x ½ = 1/16 = 6%. This means the chance of getting a tail in the fourth flip is 94%. If I observe three heads in a row, isn't it a sure bet to get a tail in the next flip?

In our experiment, we only recorded the next flip after three consecutive heads, and yet, we see the same random walk pattern for the fourth flip after three consecutive heads. The fourth flip after three consecutive heads still seems to be a random 50% chance.

The resolution of this seeming paradox is described in two ways.

First, at every flip, there is the same chance of getting heads or tails. There is no connection or correlation between one flip and the next. They are each independent. This means the chance of getting a heads in the next flip is exactly the same if there was a single tail or twenty heads prior.

Here is what is wrong in our reasoning of expecting a 94% chance of getting a tail after 3 consecutive heads.

When we do the initial three flips, we have eight different possible outcomes. Seven of these outcomes happen and we just set them aside. Only for the 1 in 8 outcomes do we go the next step and record the fourth flip.

Once we decide to record the next flip, the chance of getting a heads or a tails in the next flip is the same. We are looking at this specific path, but the other paths are happening, we are just not looking at them. This is illustrated in Figure 7.7.

*Figure 7.7. Diagram of the outcomes of sixteen coin-flips. All happen. We just look at the specific ones along the outer edge.*

When we flip one more time, we still have only two choices. Half will be heads, half will be tails. The chance of getting a tail on the next flip is still only 50%.

It's just that we are only sampling 12% of all the flips.

## 7.15 Summary of the commands introduced so far

| Command | Description |
|---|---|
| `void setup(){`<br>`}` | This is a function that appears in EVERY sketch. Every command within the brackets will be executed just once |
| `void loop(){`<br>`}` | This is a function that appears in EVERY sketch. Every command within the brackets will execute over and over again. |
| `pinMode(13, OUTPUT);` | This command tells the Arduino that we are going to use a specific digital pin as an OUTPUT, as distinct from an INPUT. We use the pin number to identify the pin we want to setup. |
| `digitalWrite(13, HIGH);` | This command controls the output of a digital pin and makes it either a HIGH (logic 1, or 5 V) or a LOW (a logic 0 or a 0 V). Once executed, the pin value will be set to this value. |
| `delay(1000);` | This command tells the Arduino to sit there, twiddling its thumbs, doing nothing for a duration of milliseconds as listed inside the (). In this example, the time interval to wait is 1000 msec = 1 sec. |

| | |
|---|---|
| `//`<br>or<br>`/*`<br>`*/` | Comments. Everything after these two forward slashes will be ignored by the Arduino sketch. We can use this add comments for our own benefit on any line.<br><br>For multiple lines add them between the /* and the */ lines. |
| `Serial.begin(9600);` | This command opens up the serial com link with a baud rated inside the (). In this example the baud rate is 9600 baud. The name `Serial` is an object, the serial link. The `.begin` is a verb, telling the object to get set up. |
| `Serial.print("Hello World");` | This command will print characters to the serial communications channel. Everything inside the () will be printed. To print a string, enclose the characters within quotes. After printing, the cursor will be left on the same line. |
| `Serial.println(" ");` | Does the same thing as the `Serial.print` command, but moves the cursor to the next line after printing the characters. This starts the next printed content on a new line. |
| Serial monitor | The terminal emulator that you can print to. This is the primary way of displaying information from the Arduino. |

## 7.15 Summary of the commands introduced so far

| | |
|---|---|
| `int, long` | This will create a new variable that will store an integer. This is a whole number. It is usually placed before the setup() function |
| `float` | This will create a new variable that will store a floating-point number. This is a number that has a decimal point. It is usually placed before the setup() function |
| `+, -, *, / =` | Simple algebra |
| `==` | A conditional test: are the two terms on either side of the == the same. Not a misprint, double equal signs. |
| `if (iVal == 1) {`<br>`}` | If conditional. If the statement inside the () is true, then the commands between the brackets are executed. |
| Tools/Autoformat | Will add indents and clean up the look of your code |
| Serial Plotter | A built-in graphing features which plots the numbers printed to serial com port. |

| | |
|---|---|
| `for (index = 1; index`<br>`<= npts; index =`<br>`index + 1) {`<br>`  //execute this code`<br>`}` | For loop. There are three statements inside the () of the for loop. The first sets the initial value of the index. The second term is the condition. If the condition is true, continue in the loop, otherwise, leave the loop. The third term is the increment. The code inside the {} is executed in the loop. |
| `random(min, max)` | Returns a random, long integer number between min and (max-1). |
| randomSeed() | This term will start the random number generator with the number inside the (). |
| while (this is true) {<br>`}` | While loop. Will execute the commands in side the brackets as long as what is in the () is true. |

# Chapter 8. Measure Static Electric Fields

The Arduino is really designed to interact with the external world by measuring voltages and creating voltages. But in most examples, we need to plug components into the Arduino that either respond to or create voltages. These are generally sensors or actuators.

In this video eBook, I purposely limited our experiments so that we did not require any additional components, just the low-cost Arduino board itself. It just means we can't do many voltage related experiments. But there is one we can do that involves nothing more than a steel paperclip.

## 8.1 What you will need, commands you should know, commands you will learn

We've introduced the skills needed to perform calculations and print results. Two key elements are the use of variables, both integer and floating-point variables, and the use of loops.

In the last chapter we introduced the *for loop* and the *while loop*. These helped shape the flow of the code.

In this chapter, we add to our collection of tools the ability to read a voltage from the analog pin. With this ability we will open a window into the invisible world of the electric fields all around us.

## 8.2 Static electric fields or 60 Hz pickup

If we could see the electric fields around us, we would see we live in a dense fog of electric fields. High frequency fluctuating fields are from our communications products- cell phones, Bluetooth, Wi-Fi, rf links between a mouse and computer, cordless phones, portable radios and radio or TV stations.

These intentional radiators broadcast at frequencies from 1 MHz to over 10 GHz. While our cell phones and computers are sensitive to these very high frequency electric fields, the Arduino is not.

But the brightest source of electric fields around us is the pollution from the power lines that distribute 110 V ac power around our home to all the appliances that plug into the wall. The power lines radiate at a frequency of 60 Hz in the US. If we could see the 60 Hz electric fields, we would see bright spots all around our house where the wiring is. It would be like we strung ropes of LEDs all around the house that were constantly on, flooding us in their 60 Hz light.

The Arduino is sensitive to low frequency electric fields. It can sense these 60 Hz polluting electric fields under special conditions. In addition, there is another source of even lower frequency electric fields, almost DC.

In the normal course of moving around, we build up excess static charge. Rubbing between the surfaces of our clothing can build up lots of excess static charge which can generate thousands of volts between different surfaces.

These static charges will generate large stray static electric fields that change very slowly compared to the 60 Hz power line electric fields. And these static electric fields can also be measured by the Arduino.

## 8.3 Turning electric fields into voltages with an antenna

The Arduino analog to digital converter (ADC) pin can only measure a voltage. We need a *sensor* to convert the electric fields into voltages. The specific sensor we will use is called an *antenna*. In this experiment, the antenna will literally be a wire we stick in the ADC pin.

Any changing electric field that hits the wire will induce currents up and down the wire, causing an alternative buildup of +++ or --- charges at the ADC pin. These excess charges will create a voltage and it's the voltage we will measure.

Unfortunately, the same features that make the Arduino more sensitive to detecting static electric fields from moving nearby static charges, also makes it more sensitive to the pollution from 60 Hz ac fields. Afterall, the 60 Hz pickup is also from stray fields. It's just that they are fluctuating at 60 cycles per second.

We distinguish the electric fields from the ac power cords by referring to them as *AC electric fields*, while the electric fields from excess charges stuck to surfaces as *static electric fields*.

To measure these stray electric fields, all we have to do is take some readings from the ADC pin of the Arduino and plot the voltage on the plotter.

## 8.4 New hardware feature: ADC

Built into every Arduino microcontroller is a special section of the chip that will measure the voltage on an input pin. This voltage is called an *analog signal*, as distinct from a digital signal.

Many of our previous experiments involved digital signals. These types of signals have only one of two values, a LOW or a HIGH, a logical 0 or a logical 1. In the Arduino Uno board, the voltage of a logical 1 signal is 5 V and the voltage of logical 0 signal is 0 V.

That's it. Those are our only choices in a digital signal.

Analog signals can have almost any value within some range. To be read directly by an Arduino, the voltage level is limited to between 0 V and 5 V, but we can measure many different values between these limits. Figure 8.1 illustrates how the voltage of a digital and analog signal might vary over time.

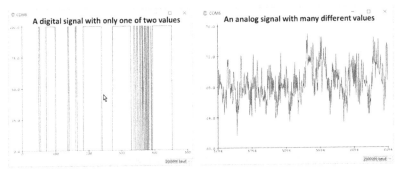

*Figure 8.1. An example of a digital signal (left) and an analog signal (right).*

The analog voltage is measured and converted into digital information to then be interpreted by the Arduino, with the *analog to digital converter* (ADC) circuit built into the Arduino microcontroller chip.

The ADC in the Arduino Uno is able to measure a voltage with 10-bit resolution. This means the voltage is read and converted into $2^{10} = 1024$ different levels, ranging in value from 0 to 1023.

When we read a voltage value with an ADC, what the Arduino gives back to us is an integer corresponding to the level between 0 and 1023. These values correspond to an input voltage of about 0 to 5 V.

When the ADC measures a voltage on one of the specific analog input pins, labeled as A0 to A5, located in the lower right side of the board shown in Figure 8.2, it is always the difference in voltage between that pin and the ground pin on the Arduino board.

Figure 8.2. The Arduino Uno with the analog pins in the lower right of the board identified by the square.

## 8.5 Taking your first ADC measurements

In the Arduino Uno, there is really just one analog to digital converter (ADC) built into the microcontroller. But, there are six different pins to which it can connect.

When we command the Arduino to take an analog reading, we are really telling the ADC which pin to connect the ADC and then take the reading. On the Atmega 328 microcontroller chip, the brains of the Arduino Uno board, there is one ADC and a switch which connects one of the six analog pins into the ADC, one at a time.

This means we can only take readings from one analog pin at a time. The switch can connect the ADC to a specific analog pin pretty quickly. It can connect and take a reading in about 112 microseconds. This is about 8,000 times a second.

In the Arduino board, all six of the analog pins that connect to the ADC are available to sensors. We identify them and select them by their analog pin number, A0, A1,...A5.

To read the voltage on any pin, we use the command,

analogRead(pinNumber);

The pin labels shown on the side of the pins on the board, are the pin numbers we use in the command to read that pin. This command will return the value read by the ADC as a number from 0 to 1023. I call these units analog to digital units, or _ADU.

Every time we call this analogRead() command, we take a reading from the specified ADC pin and store this number in the command. We could just print this value each time we read it.

To take a reading on analog pin A0, and print the value to the serial monitor, is literally one line of code:

Serial.println(analogRead(A0));

***Try this experiment***: Write a sketch that sets up the Arduino to print the value on the analog pin A0 and plots it on the serial plotter, as fast as it can. What do you think you are measuring?

The whole sketch is only six lines. Here is my version:

```
void setup() {
  Serial.begin(2000000);
}
void loop() {
  Serial.println(analogRead(A0));
}
```

## 8.5 Taking your first ADC measurements

When you run this sketch, you will plot on the serial plotter (don't forget to set the baud rate to 2000000) the voltage appearing on the A0 pin of the Arduino board.

Even with nothing connected, you will see a voltage. What I measured when I did this experiment is shown in Figure 8.3.

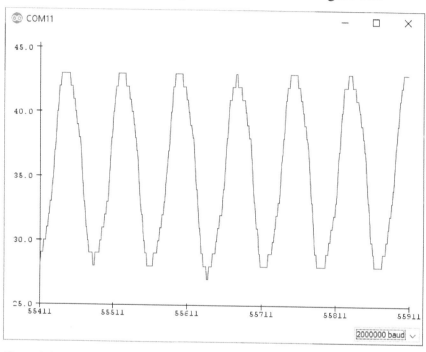

*Figure 8.3. The voltage I measured on analog pin A0 with nothing connected and plotted on the serial plotter. Note the scale is in ADU from 0 to 1023*

What are we measuring?

Congratulations, you are measuring the 60 Hz pick up from stray electric fields in the room around you, radiating from all the power lines nearby. The power line voltage has a frequency of 60 Hz. This means there are 60 cycles of the voltage every second. The period of each complete cycle is 1/60 Hz or 16.67 msec. This is a good number to remember. The sine wave we see in this measurement has a period of 16.67 msec.

This is the pollution noise that will keep us from seeing the lower frequency static electric fields from static charges. We will need to do two important things to minimize the 60 Hz pollution.

The first and most important thing is to move your Arduino away from any nearby power cords. The power cords and the wiring inside your walls are like bright lights shining on your Arduino. Turn off any lights that might be nearby as these will have 60 Hz currents and contribute to the 60 Hz pollution noise.

***Try this experiment***: make a note of the peak to peak value of the waves on your screen. Then move a power cord close to your Arduino. By how much does the peak to peak value increase?

Once you isolate your Arduino as best as you can, the second important technique we will apply is averaging out the 60 Hz noise.

> *Watch this video and I will show you the sensitivity of my Arduino to stray 60 Hz fields.*

## 8.6 Storing ADC values in a variable in units of ADU

The analogRead () command returns a reading from the ADC channel. In the previous experiments, we just printed this out. We can also put this value in a variable and do more analysis with it.

While the level or count value from the ADC is just an integer and dimensionless, to remember that these integers came from an ADC and we are counting ADC levels, I add to the end of the variable

## 8.6 Storing ADC values in a variable in units of ADU

name the generally accepted unit label of *Analog to Digital Units*, or ADUs.

For example, a variable storing an ADC value might be:

iADC_A0_ADU = analogRead(A0);

The way I decode the variable is:

- *The i means this variable is an integer*
- *The ADC_A0 means this variable is measuring what appears on the ADC A0 pin.*
- *The ADU means the units of this variable are analog to digital units, varying from 0 to 1023.*

Using variable names, we can write code that is a little more flexible. If we use variable names that are self-documenting, we can literally read the code and decipher what it does without needing a bunch of comment lines.

Should a reading from the ADC be an integer or a floating-point number? The difference is an integer has no decimal point, but a floating-point number does.

When we are just going to use the raw ADU value from the ADC, using an integer value is just fine. But, if we plan to do any averaging or more complex analysis, using integers will round-off any fractions and we would lose a little accuracy.

> *If we plan to do any calculations with the ADU values, we should use the variable as a floating-point number. This will give us more accuracy.*

For example, if we use an long type variable to store the ADU value and measure the average of 1000 readings the average value will be rounded down to an integer value. We lose the fractional information.

If the variable that stores the ADU values is a float type, we retain the fractional value of the average. This gives us a little more precision.

For this experiment measuring electric fields we really don't care what the voltage value is at the pin. Using the ADU value and displaying this is just fine for us.

***Try this experiment:*** Read the voltage from the analog pin, place it into a variable and print the variable. Use a delay so you print the voltage to the plotter every 10 or 20 msec.

Here is my sketch:

```
float V_ADU;
int iDelay_msec = 20;
int pinADC = A0;
void setup() {
  Serial.begin(2000000);
}
void loop() {
  V_ADU = analogRead(pinADC) * 1.0;
  Serial.println(V_ADU);
  delay(iDelay_msec);
}
```

We are controlling the plotting time base with the delay line. With 20 msec between measurements and 500 measurements full scale, this is 0.02 sec x 500 = 10 seconds full scale. This is a convenient time scale to watch the data scroll by.

What do you see? My initial measurement is shown in Figure 8.4.

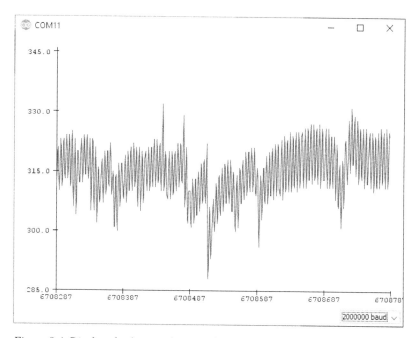

*Figure 8.4. Displayed voltage with a sampling interval of 20 msec. During the 10 second period displayed on this screen, I moved my hand closer and farther to the A0 ADC pin causing the voltage to change in the sharp dips.*

The time between data points is spaced 20 msec. The 60 Hz power line noise has a period of 1/60 = 16.67 msec. Even though we are sampling slower than the 60 Hz voltage pattern, we are still seeing a large amplitude of fluctuating voltage plotted.

Just taking data at a slower rate than the 60 Hz power line frequency will not remove the 60 Hz noise. Instead, we are going to take advantage of a powerful trick.

## 8.7 Averaging over n power line cycles (PLC)

The 60 Hz voltage pattern has just as many voltage values above the average as below. If we were to average all the voltage readings over one 60 Hz cycle time, with the same pattern above 0 as below 0, we should average to nearly 0. The more cycles we average, as long as it is over a whole number of complete cycles, the closer to the 0 average we will get.

Of course, the more cycles we average over, the longer the time between average values and the slower the plotting. There is a tradeoff.

We are going to write the code to take measurements as fast as we can. Then we are going to average all the voltages we can take over n power line cycles, each lasting 16.67 msec. It doesn't matter when we start as long as we stop averaging exactly 16.67 msec, or an integer multiple, after we start.

This process is called digital filtering. It is a powerful technique to eliminate periodic noise, used in many communications products.

After we implement our digital filter the remaining voltage should be due mostly to the influence from static electric fields.

To perform the average over a fixed time we are going to introduce two new functions, the *while loop* and the *microsecond timer*. Measuring time in microseconds will give us a little better resolution to average over the 16.67 msec period than using units of msec.

The syntax of the while loop function is simple:

```
while (x < 10) {
  //Execute this code
}
```

The condition inside the () is evaluated each time the while loop starts. If the condition is TRUE, then the commands in the while loop execute. When the condition is false, execution flows to the point after the closing }.

When any sketch begins in the Arduino, it automatically starts two counters counting up from 0. The millis() counter counts the number of milliseconds that has passed since the start of the sketch. The micros() timer counts the number of microseconds that have elapsed since the start of the sketch. The functions millis() and micros() have the current count values. The counts in milliseconds or microseconds are stored as *long* integers.

If we want to save the current count of microseconds, we assign the micros() value to a variable. In this example I created a variable in the sketch called iTime0_usec:

iTime0_usec = micros();

If we want to measure an elapsed time, we store the start time from the counter, and compare the current micros() time to the stored time. The difference is the elapsed time.

***Try this experiment***. You know enough now to think through the details for the algorithm you would want to use to average over n power line cycles, print the average and take another point. Take a moment to write down an algorithm you would use and try your hand at the sketch.

When you are ready check out my sketch in the next section

## 8.8 My sketch: display measurements averaged over n power line cycles

```
// Measure static electric fields for n PLC
int pinADC = A0;
int nPLC = 1;
long iTime2Average_usec = (1000000.0 * nPLC )/ 60.0; // time we
want to average
float V_ADU;
long nCountsActual;// number of actual measurements averaged
long iTimeStart_usec; // start of averaging time
long iTime2Stop_usec; // stop of averaging time
/////////////////////////////////////////////
void setup() {
  Serial.begin(2000000);
  for (int i = 1; i < 3000; i++) {
    V_ADU =analogRead(pinADC);
```

```
  }
}
void loop() {
  ////initialize variables at start of loop///
  V_ADU = 0.0;
  nCountsActual = 0;
  iTimeStart_usec = micros();
  iTime2Stop_usec = iTimeStart_usec + iTime2Average_usec;
  ////////////////////////////////////////////
  while (micros() < iTime2Stop_usec) {
    V_ADU = V_ADU + analogRead(pinADC) * 1.0;
    nCountsActual++;
  }
  V_ADU = V_ADU / nCountsActual;
  Serial.print(nCountsActual); Serial.print(", ");
  Serial.println(V_ADU);

  ////plot with fixed scales
  //Serial.print(V_ADU);      Serial.print(", ");
  //Serial.print(0); Serial.print(", ");
  //Serial.println(100);
}
```

*Watch this video and I will walk you through this sketch.*

I used three tricks in this sketch that may not be obvious.

When we start using an ADC channel, in the way we are, with nothing connected, there is some initial charge stored on the input of the ADC that gives some initial funny values; a start-up transient *artifact*. After taking a bunch of readings, the ADC has eliminated this residual charge and the artifact is eliminated.

***First***, to get rid of this artifact before we actually take the measurements we want, I added some dummy steps in the setup() function. I told the Arduino to make 3,000 ADC measurements and just throw the values away. This cleans out the residual artifact. Here are those lines:

## 8.8 My sketch: display measurements averaged over n power line cycles

```
for (int i = 1; i < 3000; i++) {
   V_ADU =analogRead(pinADC);
}
```

You might have seen these glitches in your measurements. Figure 8.5 shows the plotter measurements if I do not have this initial clearing out process, and then adding it in. The difference is remarkable.

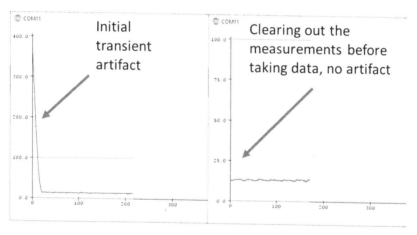

Figure 8.5. Adding the for loop in the setup() function clears the data and there is no initial transient artifact.

**Secondly**, I wanted to be able to see how many measurements I took in each average. I was curious, when we take data as fast as we can, how many data points are we averaging in one or 2 power line cycles.

When I average over 1 power line cycle, and take data at the rate of about 8,000 measurements per second, I expected about 8,000 measurements/second x 16.7 msec/cycle = 133 measurements averaged together per cycle.

I added the print statements to print the nCountsActual and the V_ADU values:

```
Serial.print(nCountsActual); Serial.print(", ");
Serial.println(V_ADU);
```

When I ran this code, I saw on the serial monitor that I was taking about 132 measurements in 1 power line cycle. This is very close to what I expected and is an important consistency test. It gives me a little more confidence I understand what the code is doing.

Once I know this value, I comment these lines out and just print values I want to plot.

***Third***, if all I did was plot the voltage values on the plotter, the serial plotter would continually auto scale the plot for me. When we use this to show the impact from static charges, I don't want the scale to change, I want the scale to be fixed so I can see the relative static fields from different sources.

I added two special print lines to force the serial plotter to auto scale on these values. As long as my data to plot is within these limits, the serial plotter scale will be fixed and not change. Here are those lines:

```
////plot with fixed scales
  //Serial.print(V_ADU);      Serial.print(", ");
  //Serial.print(0); Serial.print(", ");
  //Serial.println(100);
```

After I see the count and ADU values on the serial monitor, I comment out those print lines and uncomment these three print lines. I can then switch to the serial plotter and see the voltage on the pin, now with the 60 Hz pick up dramatically reduced. The ADU counts at this point are shown in Figure 8.6.

## 8.9 Add an antenna to increase the sensitivity to stray electric fields

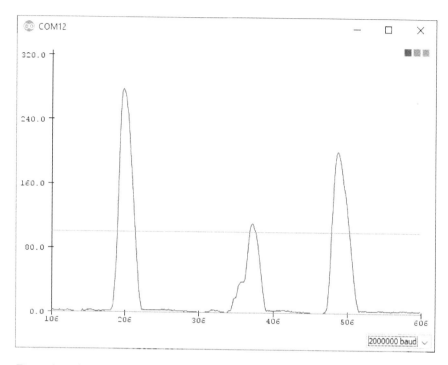

*Figure 8.6. The measured ADU values on the ADC channel with 1 power line average and a fixed scale of 0 to 100 ADU full scale. The peaks are from me moving closer or farther away. I am charged up from moving in my chair.*

The good news is that we do not see any 60 Hz pickup. And we are sensitive to static electric fields. I charged myself up a little and moved closer to the Arduino and father away, generating the peaks we see there. The bad news is we are not very sensitive to stray fields. We can improve this with a better antenna.

## 8.9   Add an antenna to increase the sensitivity to stray electric fields

So far, every experiment we've done in this video eBook has involved just the Arduino board, nothing else. In this experiment, we need to add one small component in order to increase the sensitivity of the ADC pin to static electric fields. The antenna we will add is basically a wire. Any wire a few inches long will do. If you don't have a handy wire, a paper clip will do.

There are usually three types of popular paperclips you may have lying around. Only one of them will work. Figure 8.7 shows examples of these three different types of clips.

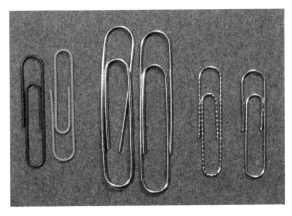

*Figure 8.7. Three different types of common paperclips. The left ones are plastic. The middle ones are too big. The ones on the right will work.*

The colored paperclips are plastic. These are insulators and won't work to make a connection to the Arduino pins.

The ones in the middle are too large a wire diameter and won't fit in the Arduino header pin holes.

The ones on the right are just barely usable. Thinner wire would be better, but these will do. The ideal wire diameter is 25 mil diameter, which, is 22 AWG gauge. You can view the table of wire diameters and AWG gauge here, for example.

The typical paperclip diameter is 34 mil diameter, or closer to 20 AWG gauge. With a little effort it can be pushed into the holes of the headers.

You want to use the thinnest, most flexible metal paperclip you have lying around.

Unwind the paper clip and let it stick up in the air. Any stray electric fields nearby will induce voltages in the wire which will be measured by the ADC. The longer the wire, the higher the voltage induced for the same electric fields. But the longer the wire, the larger the 60 Hz pick up as well.

After you insert the paperclip in the ADC pin, run the sketch to plot the static field again. You will see much larger peaks now. With this tool, we are ready to explore static charges.

## 8.10 Measure static charges

While static charges are built up all around us, one of the simplest ways of getting a large buildup of charge is to just unroll a little packing tape of any sort. Figure 8.8 shows a piece of tape next to the roll, close to the ADC antenna. This is all that is required to make huge static electric fields.

*Figure 8.8. Unrolling a piece of tape generates huge excess charges which are easily detected by the antenna in the A0 ADC pin.*

When I move the tape closer or farther from the antenna, I can see the ADC voltage increase as I move the tape. This is due to the influence of the proximity of the static charges on the antenna stuck in pin A0.

Figure 8.9 shows an example of the measured voltage on the ADC when I moved a piece of freshly unrolled of tape closer and farther from the antenna.

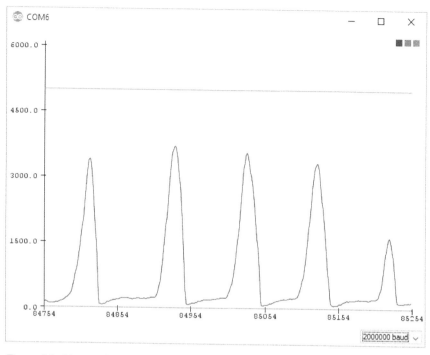

*Figure 8.9. Measured voltage on A0 as I moved a piece of tape closer, then farther away.*

When I moved the tape closer, the voltage increased. But if I just left the tape in position and did not move it, the voltage dropped pretty fast, all by itself. Then when I moved it away, it dipped slightly, and repeated again every time I moved the tape closer.

*Watch this video and I will demonstrate measuring static electric fields from static charges create by triboelectricity*

What is going on? Why does the ADC respond this way? Clearly I am measuring the influence of static charge, but why does it change this way?

## 8.11 How the ADC measures electric field

Part of the measurement process in any experiment is thinking through what exactly is being measured. This is called *situational awareness*.

Fundamentally, what is really measured by an ADC is a voltage. How is the presence of static charge converted into a voltage to measure?

To think about this, we are going to look at three simple models of the ADC with increasing complexity.

As a starting place to analyze how an ADC converts a static electric field into a voltage, we will approximate the ADC as an ideal amplifier that just measures the voltage at its front and converts this into a digital signal. It has a big resistor at its front. The value of this resistor is very large, like on the order of 100 Meg Ohms. A current through this resistor generates a voltage and this is measured by the ADC. This is illustrated in Figure 8.10.

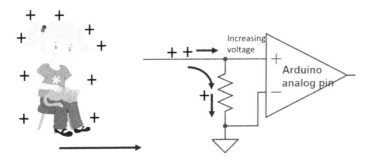

*Figure 8.10. When nearby static charges move closer to the antenna, same sign currents in the antenna are pushed through the internal, large resistor of the ADC which creates the voltage we measure.*

When we rub a balloon in our hair or over a sweater, the balloon gets charged up, usually negatively. It has a lot of excess negative charges on it.

### 8.11 How the ADC measures electric field

When this charged balloon approaches a conductor, which has some charges free to move, the charged balloon will push same-sign-charges away from it. Figure 8.11 illustrates this.

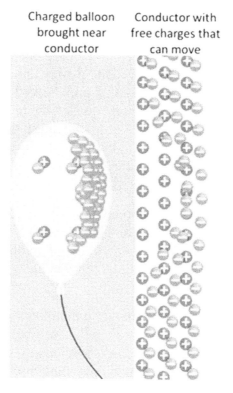

*Figure 8.11. Negative charges on the insulating balloon are stuck and do not move. When it is brought close to a conductor, it repels negative charges into the conductor.*

When we push the balloon closer, the freely moving negative charges in the conductor are pushed away into the conductor. When we move the charged balloon away from the conductor, these free negative charges flow back.

This is what is going on in the antenna:

1. *The charged balloon induces a current in the antenna sticking out of the ADC pin, due to the balloon's excess charge and its motion.*

2. *This current in the antenna flows through the large resistor in front of the ADC amplifier generating a voltage across the resistor.*

3. *This voltage is measured by the ADC.*

In this simple view, when I moved the freshly unrolled tape closer to the antenna, the excess +++ charge on it moved closer to the antenna. These excess +++ charges repelled the plus charges in the antenna and pushed a little bit of positive current through the large resistor at the front of the ADC. This positive current created a positive voltage across the resistor and was read as a positive voltage, for a short time, by the ADC.

When the static charges just sit there motionless in front of the antenna, they are not pushing charges in the antenna and the current through the resistor is 0 and there is no voltage measured. It is only <u>changes</u> in the position of the static charges that are measured.

When the external +++ static charges are pulled away, the plus charges that had been pushed away in the antenna, flow back and the current direction through the resistor changes, so the voltage polarity across it changes. Normally, we should see a negative voltage read by the ADC when we pull the +++ charges away.

But, in addition to the large resistor across the ADC, there is also a diode that prevents any negative voltage from appearing on the front of the ADC. We can only measure positive currents through the resistor.

In addition, there is one other circuit element across the ADC.

If there were only a resistor across the ADC then the instant the static charges stopped moving, the current would stop and the measured voltage would immediately go to 0 V.

But this is not what we see. When you suddenly stop moving the static charges, the ADC voltage does not immediately drop to 0, it drops over a short time interval, a fraction of a second.

The spec sheet for the Arduino Atmega 328 ADC shows a 14 pF capacitor at the input to the ADC. A slightly more advanced, second order, model for the input to the ADC includes this capacitor and the diode to show that only large positive voltage changes will appear on the ADC. This second order model is illustrated in Figure 8.12.

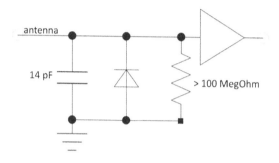

*Figure 8.12. A second order model of the ADC including the diode and small capacitor.*

The combination of this capacitor and the large resistor means that if we move the external charge close to the antenna quickly, the induced charge in the antenna moves into the capacitor to charge it up. If the external static charge stops moving, the charge on the capacitor discharges through the large resistor.

The time for the capacitor to discharge, and the voltage to drop to zero is related to the value of the resistor and the capacitor. This model helps to understand the general features of the detector:

1. The ADC only measures _motion_ of the static charge.
2. If the static charges stop moving, the ADC reads 0 V.
3. If a plus charge moves closer, we measure a positive voltage.

4. *If a plus charge moves away, we would measure a negative voltage but the diode keeps the voltage from dropping below 0 V.*

5. *If a negative charge moves closer, negative charges would be pushed through the resistor, but the diode prevents us from measuring a negative voltage.*

6. *If a negative charge moves away, we measure a positive current in the resistor.*

All we can measure is a positive voltage. This is how we can use these principles to measure the sign of the static charge:

- *If we move the charge <u>closer</u> and we see a <u>positive</u> voltage change in the ADC, it's a positive charge we are moving closer.*

- *If we move the charge <u>away</u> and it is a <u>positive</u> voltage, the excess static charge on the object is negative.*

## 8.12 A simple static charge experiment with a cat

Static charges are all around us. We cannot prevent static charges from building up on insulators.

In a humid climate, on every surface is a thin film of water. This makes the surface a little conductive, which will allow excess static charges to flow around and equalize. It's harder to build up static charges when its humid.

The opposite is also true. In a dry climate, most surfaces are insulating, and static charges can build up creating very high voltages. In fact, the voltages built up from static charges can reach thousands of volts, high enough to breakdown air. This is why you will sometimes get a spark in air after walking on a carpet or petting a cat and touching something grounded.

Usually, when any two insulating surfaces rub over each other, static charges are built up through a process called *triboelectricity*. When you rub a balloon on your sweater or in your hair, the balloon picks up static charges.

A notorious source of static charge is cat fur. Petting a cat leaves you each charged up.

With the assistance of my lab-cat, Schrodinger, I had a convenient source of static charge right next to my Arduino. Figure 8.13 shows Schrodinger in position to offer static charge whenever I pet him. Even though he happened to be resting on a static dissipation mat, he was powerful enough to still provide significant charge.

*Figure 8.13. Schrodinger, my lab-cat, working hard to provide a local source of static charge.*

After a few pets, I was charged up. Moving my hand close to the antenna, and then quickly away, resulted in large positive peaks in voltage, only when I moved away. This was very repeatable. Figure 8.14 is an example of a series of motions, moving closer slowly and pulling away quickly.

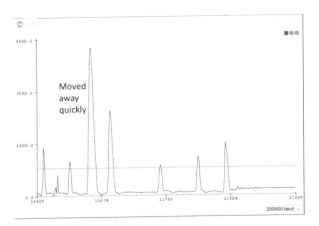

*Figure 8.14. Measured response of moving my hand away from the antenna after having petted a cat and getting charged negatively.*

In this experiment, I measured a positive voltage when I moved my hand away from the antenna. This means that I was charged negatively. When I moved forward, I induced a negative current in the resistor of the ADC, which the ADC did not measure because the negative current went through the diode, with a small negative voltage which could not be measured by the ADC.

When I pulled away, I allowed the negative charges to move back into the antenna, now generating a positive voltage across the large resistor connected across the ADC.

## 8.13 Static charge experiments

The combination of the antenna sticking out of the ADC pin and the sketch we wrote, is a static charge detector. By measuring whether we got a positive voltage by moving closer (positive charge) or moving away (negative charge), we can measure the sign of the charge on the objects moved closer and farther away.

Just as with sniffing 60 Hz stray fields, which opened up a window into the invisible world of ac electric fields all around us, we now

## 8.13 Static charge experiments

have a window opened up to sniff the static electric fields all around us.

*Here are some experiments to try:*

Two really common sources of static charge are unrolling scotch tape or packing tape and pulling apart two pieces of Velcro. If you take either one of these examples, there is a good chance it will be very charged.

Take one of these objects and move it closer or farther from the antenna and see if it can detect you. Think about what is actually being measured and why you see this signature. For each object you move closer, it is positively or negatively charged?

Rub your hand over a shirt or sweater and bring it closer and farther from the Arduino. What is the sign of the charges on your hand? Try other surfaces to find a positive charge, and a negative charge source.

If you do not move, you should see the voltage on the serial plotter not change. This should convince you that it is *changes* in the induced charge, flowing through the 100 MegOhm resistor that we are really measuring. Don't move and there is no change in the induced voltage.

When we sit in a chair, when we move around, when we rub our arms, we build up a net, excess, static charge by *triboelectricity*.

When different types of insulating surfaces rub each other, opposite charges are pulled apart and one surface gets charged negatively and the other positively.

We see a striking example of this when we walk along a rug on a dry day and touch a piece of metal, seeing a large spark. We get charged up positively and leave a trail of negative charges on the rug behind us. Depending on what type of clothes we are wearing, we can charge up with an excess negative or positive charge.

It is incredibly difficult to NOT have an excess charge. We have to wear special clothes that are slightly conductive, and our body needs to be connected to a grounded outlet with a wire. It also helps if the room humidity is greater than 50%. Not likely in Colorado where I live.

Here are some other great sources of static charge. Generally, when one surface gets charged negatively the other surface gets charged positively. You can confirm this in the following materials:

1. *scotch tape or other tape*
2. *clear plastic cellophane wrapping, the more "clingy" the better*
3. *two different fabrics, like wool and nylon, or cotton, or polyester rubbing against each other*
4. *your hair and a rubber balloon*
5. *your hand and a cat*
6. *rubbing a wool sweater*
7. *sliding across a cloth chair*
8. *sliding across a plastic chair*

When dealing with sensitive electronic parts, your static charges can potentially blow them up. After you practice building up static charges, practice reducing your static charge build up.

How can you reduce the static charge you carry around? You can use your static charge sniffer to first charge yourself up and then see how effective you can be discharging yourself.

Try some of these methods to reduce your static charge build up:

1. *Keep the room humidity > 50%. Above this humidity, most surfaces have a very thin layer of water condensed on*

them. *This layer is slightly conductive and bleeds off static charge to any other conductive surface.*

2. *Wear a metal wrist strap that is connected to a good earth ground to bleed off some of your static charge*

3. *Use a tabletop conductive pad, called an ESD mat, connected to a good earth ground to keep static charges from accumulating on the table.*

4. *Wear special clothes which do not have a strong triboelectric effect.*

5. *Rub a wire connected to ground over your arms and clothes to draw off excess charges.*

Can you make yourself invisible to the Arduino electric field detector?

Unless some of these precautions are taken, everyone will always have a net charge. This charge will generate an electric field around us. We can't see or feel this electric field, but we can use the Arduino analog pin to measure this invisible presence for us.

## 8.14 Electrostatic damage (ESD)

The static charge we build up just due to our normal activities can be disastrous to some electronic components that are not protected. When we are charged up and then touch an electronic component and our charge flows through the component, the large electric field and current burst can damage or destroy the component.

We call what happens when our charge bleeds off onto another device, electrostatic discharge, ESD. When this causes a part failure, we call it ESD damage.

Cat fur has a notoriously high triboelectric effect. As cats move, they build a large static charge. If we pet a cat, we get highly charged.

My other cat, Maxwell, is a great ESD tester. As he walks around my lab bench, touching his nose to my circuit boards, he often delivers a large ESD pulse to the electronics. Parts which are internally well protected with diodes survive. Figure 8.15 shows Maxwell hard at work testing the ESD sensitivity of an Arduino Uno board. One reason I like using the Uno board is that it is very robust to all of Maxwell's ESD tests.

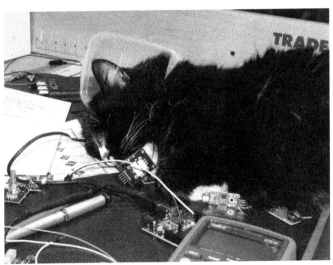

*Figure 8.15. My lab-cat, Maxwell, is a very good ESD tester. His fur builds up a large static voltage which he discharges through his nose when he touches electronic components. The Arduino Uno is very robust to ESD damage which is why I like to use it for my workshops.*

## 8.15 An advanced note

The second order model we created for the front of the ADC is a pretty good model which will help us interpret the measurements we see and how the ADC responds to extra static charges. But, it is not exactly what is going on in the front of the ADC.

In our simple model, we assumed there was a 100 Meg Ohm across the front of the ideal voltmeter of the ADC. In most situations, it's really okay to just think it is 100 Meg Ohms. But in reality, it's not really a resistor. It is a current source created as part of the transistors that make up the ADC circuit.

A better model, a third order model, to describe the front of the ADC is the circuit shown in Figure 8.16.

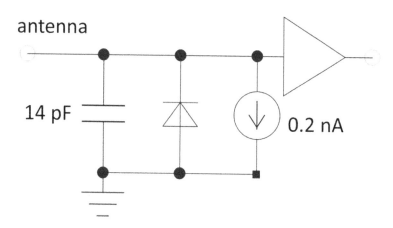

*Figure 8.16. A third order model of the ADC using a current source.*

The current source is the input bias current for the amplifier that feeds the ADC. In its normal operation, it draws a steady 0.2 nA or 200 pA of current. If we charge up the 14 pF capacitor by moving some static charge in front of the antenna, this current source will discharge the capacitor in a time of about

$$t = \frac{CV}{I} = \frac{14\,\text{pF} \times 5\,\text{V}}{0.2\,\text{nA}} = 0.4\,\text{sec}$$

The current source will bleed the charge from the capacitor linearly in time. This is why the drop off in voltage after I stop moving the charge is more of a linear drop in about 0.4 seconds, rather than the decaying exponential we would expect from an RC circuit.

The third order model is not important to know when using the Arduino ADC to measure static charges. Our simple second order model explains the effects we see with static charges.

## 8.16 Summary of the commands introduced so far

| Command | Description |
|---|---|
| `void setup (){`<br>`}` | This is a function that appears in EVERY sketch. Every command within the brackets will be executed just once |
| `void loop (){`<br>`}` | This is a function that appears in EVERY sketch. Every command within the brackets will execute over and over again. |
| `pinMode(13, OUTPUT);` | This command tells the Arduino that we are going to use a specific digital pin as an `OUTPUT`, as distinct from an `INPUT`. We use the pin number to identify the pin we want to setup. |

## 8.16 Summary of the commands introduced so far

| `digitalWrite(13, HIGH);` | This command controls the output of a digital pin and makes it either a `HIGH` (logic 1, or 5 V) or a `LOW` (a logic 0 or a 0 V). Once executed, the pin value will be set to this value. |
|---|---|
| `delay(1000);` | This command tells the Arduino to sit there, twiddling its thumbs, doing nothing for a duration of milliseconds as listed inside the (). In this example, the time interval to wait is 1000 msec = 1 sec. |
| `//` or `/*` `*/` | Comments. Everything after these two forward slashes will be ignored by the Arduino sketch. We can use this add comments for our own benefit on any line. For multiple lines add them between the /* and the */ lines. |
| `Serial.begin(9600);` | This command opens up the serial com link with a baud rated inside the (). In this example the baud rate is 9600 baud. The name `Serial` is an object, the serial link. The `.begin` is a verb, telling the object to get set up. |

| `Serial.print("Hello World");` | This command will print characters to the serial communications channel. Everything inside the () will be printed. To print a string, enclose the characters within quotes. After printing, the cursor will be left on the same line. |
|---|---|
| `Serial.println(" ");` | Does the same thing as the `Serial.print` command, but moves the cursor to the next line after printing the characters. This starts the next printed content on a new line. |
| Serial monitor | The terminal emulator that you can print to. This is the primary way of displaying information from the Arduino. |
| `int` | This will create a new variable that will store an integer. This is a whole number. It is usually placed before the setup() function |
| `float` | This will create a new variable that will store a floating-point number. This is a number that has a decimal point. It is usually placed before the setup() function |
| `+, -, *, / =` | Simple algebra |
| `==` | A conditional test: are the two terms on either side of the == the same. Not a misprint, double equal signs. |

## 8.16 Summary of the commands introduced so far

| | |
|---|---|
| `if (iVal == 1) {`<br>`}` | If conditional. If the statement inside the () is true, then the commands between the brackets are executed. |
| Tools/Autoformat | Will add indents and clean up the look of your code |
| Serial Plotter | A built-in graphing features which plots the numbers printed to serial com port. |
| `for (index = 1; index <= npts; index = index + 1) {`<br>`   //execute this code`<br>`}` | For loop. There are three statements inside the () of the for loop. The first sets the initial value of the index. The second term is the condition. If the condition is true, continue in the loop, otherwise, leave the loop. The third term is the increment. The code inside the {} is executed in the loop. |
| `random(min, max)` | Returns a random, long integer number between min and (max-1). |
| randomSeed() | This term will start the random number generator with the number inside the (). |
| analogRead(pinNumber); | The command to read the voltage, in units of ADU, from 0 to 1023, from the analog pin, pinNumber. |
| `while (x < 10) {`<br>`   //Execute this code`<br>`}` | While loop. Another type of loop which will repeat everything inside the {} as long as the statement inside the () is true. |

# Chapter 9. Other examples: exploring other sketches

We wrote our version of Blink in less than 10 lines of code. But we weren't the first. This is such a famous sketch that it is one of the built-in examples included with the IDE. There are hundreds of sketches available to you at your fingertips built into the IDE. In addition, there are thousands more you can download online. In this chapter we look at some of the places you can find them and how to open them up.

## 9.1 What you need to know and what you will learn in this experiment

This is the concluding chapter to Arduinos Without Tears. By this time you should be able to write a sketch that will control the onboard LED, generate variables, calculate a series of numbers using the built in loops; a for loop and a while loop.

You should be able to take advantage of advanced math features like random numbers and calculating square roots. All these series of the numbers you can print as numbers in the Serial Monitor, copy and paste them into excel, or plot them locally on the serial plotter.

These skills will be used over and over again in all future sketches you create.

As you open up and explore other sketches you will recognize these commands.

When you look at other sketches, be aware that just because the sketch is published, does not mean it was well written and is an example of how you should write a sketch. Use your new-found judgement.

## 9.2 Built-in examples

There are more than a hundred example sketches that come pre-loaded in your IDE.

Access the built-in examples from the *File/Examples* menu. This is shown in Figure 9.1.

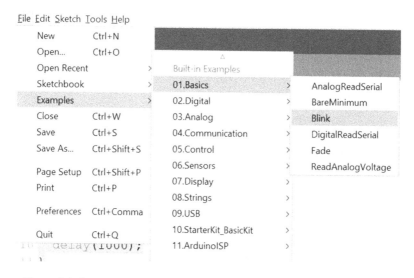

Figure 9.1. *Open and browse through some of the more than 100 examples included in the Arduino IDE.*

The examples are grouped into categories. At your leisure, you can explore all the different categories. For now, we will open the Blink sketch found in the *01 Basics category*. The sketch will open in a new window.

The canned Blink sketch is long. Browse through it. You should recognize some of the lines. Figure 9.2 shows a copy of part of the blink sketch in the examples folder.

```
20    This example code is in the public domain.
21
22    http://www.arduino.cc/en/Tutorial/Blink
23  */
24
25 // the setup function runs once when you press reset or power the board
26 void setup() {
27    // initialize digital pin LED_BUILTIN as an output.
28    pinMode(LED_BUILTIN, OUTPUT);
29 }
30
31 // the loop function runs over and over again forever
32 void loop() {
33    digitalWrite(LED_BUILTIN, HIGH);   // turn the LED on (HIGH is the voltage level)
34    delay(1000);                       // wait for a second
35    digitalWrite(LED_BUILTIN, LOW);    // turn the LED off by making the voltage LOW
36    delay(1000);                       // wait for a second
37 }
```

*Figure 9.2. Part of the canned Blink sketch. You should recognize all of these commands.*

Most of the beginning is commented out as just a description of the example.

I find in my workshops that starting out with this sketch before you have written your own is sometimes very confusing. All the comments are distracting, and the use of some special constants makes the sketch a little hard to initially understand.

In the sketch, Line 26 is where the action starts.

In the sketch, Line 28 has the pinMode command, but it is written different than expected. Instead of having 13 as the pin number, there is the word, LED_BUILTIN and it is in blue. This indicates it is a special name.

In the Arduino Uno, the built-in LED that is on the board, is always connected to pin 13. But, on other boards, like the Teensy, the built-in LED is not on pin 13, but on pin 5.

To make this program work seamlessly with other boards, the Arduino language has a built-in constant name called LED_BUILTIN.

The value of this variable changes depending on if the board is an Uno or some other board.

The rest of the sketch is identical to the one we created from scratch.

Open up and browse a few other built-in sketches at your leisure.

## 9.3   Six popular web sites with many sketches

If the hundreds of sketches in the example files are not enough, you can find thousands more online, most free to download and use.

Of course, just because a sketch has been published on-line and you can download it does not mean that it will work. Even if it works, there is no guarantee it is well written with good habits and styles you will want to copy. As with everything online, the principle to follow is "buyer beware."

Some web sites are a little more reliable than others. If you are looking for project ideas or have an idea and want to see if there is a sketch that implements it, I recommend you check these six web sites first:

Sparkfun: www.SparkFun.com

Adafruit: www.AdaFruit.com

Arduino.cc: www.Arduino.cc

Instructables: https://www.instructables.com

Hackster: https://www.hackster.io

Maker: https://maker.pro/arduino/projects

If you are searching for your next project, these are really great places to find inspiration.

# Chapter 10.  Your Next Steps

This brings us to the conclusion of this video eBook, Arduinos Without Tears.

My hidden agenda, which I tried not to hide too well, is that Arduinos are incredibly powerful building block elements upon which to launch just about any sort of electronics project you want.

If you start out developing good habits, you will develop a style which accelerates you up the learning curve and enables you to tackle ever bigger and more complex projects.

Getting started with Arduinos is not a lot of magic, but there is some jargon. The biggest three hurdles I've tried to reduce for you are

- ✓ *not having to spend a lot of money to get started with Arduinos*
- ✓ *not having to invest a lot of time to take one for a spin*
- ✓ *not having to be an electrical engineer to jump in and enjoy physical computing*

But we have just scratched the surface of what can be done with Arduinos. My own interest is using the Arduino as my lab assistant, my *Igor*. Arduinos have become my trusty lab assistant in just about every experiment and project I do in my lab.

In the other video eBooks in this series, I share with you some of the tricks I use to leverage the power of Arduinos and low-cost electronics to do, in my garage in an afternoon for a few bucks, what it used to take a year of prep and thousands of dollars in a research environment.

Check out my other video eBooks in this series: *Saturday Afternoon Low-Cost Electronics Projects*.

**Book 1**: Arduinos without Tears: the ultimate, simplest getting started guide for Arduinos with six experiments in science and coding using nothing more than a low-cost Arduino.

**Book 2**: Science Experiments with Arduinos Using a Multifunction Board, exploring more than a dozen experiments using the Arduino and a simple add on multi-function board costing about $3. No wiring needed, just plug and play.

**Book 3**: More Science Experiments with Arduinos Using an LED Accessory Kit: exploring another dozen experiments in science and electronics using the Arduino and a simple kit of parts costing less than $3.

**Book 4**: Science Experiments with the SparkFun RedBoard Turbo, exploring more than a dozen experiments using the Sparkfun RedBoard Turbo, a high-performance data acquisition Arduino.

**Book 5**: Three Must Have Instruments for Every Lab: the missing manuals for the Arduino as a voltage source and datalogger, the Digital Multi-Meter (DMM), and the scope/function generator.

**Book 6**: Electronics You Should Have Learned in High School. Starting with the basics and establishing a firm foundation in current, voltage, resistance and power, simple circuits, voltage sources, I-V curves, diodes, schematics, measuring voltage, current, resistance, voltage dividers, capacitors and inductors.

# Chapter 11. An Introduction to This Book Series.

This book is the first book in the *Saturday Afternoon Low-Cost Electronics Projects* series of books.

This series of books encapsulates much of what I have learned about computing, electronics and physics that I think is important for any hacker, maker, hobbyist, engineer-scientist or engineer-scientist wannabe. You will use these principles and skills in every future project you pursue.

This series of books is different from any of the other project books out there. All of my books

1. ...use an Arduino as the heart of the experiment

2. ... use the Arduino IDE without needing any other programming languages or skills.

3. ...require no more coding skills that you will learn in each successive book.

4. ... encourage you to create the sketches for each experiment and project.

5. ... also include my version of the sketches which you can literally copy out of the ebook and paste into the IDE and they just run.

6. ...are a "code delivery vehicle".

7. ...require very little wiring. The focus is on using off-the-shelf electronic components that plug directly into the Arduino and just work.

8. ...use just a few low-cost electronic components each costing only a few dollars. This means the more than one dozen experiments and projects in each book, providing months of Saturdays' activities, costs just a few dollars of investment.

9. ...accelerate you up the learning curve and provide valuable skills you will use in all future projects as a hobbyist, as a maker, as a tinkerer, as a DIYer, as a student, and as a professional engineer or scientist.

## 11.1 The name of this series of books says it all.

The series of books is the **_Saturday Afternoon Low-Cost Electronics Projects_** series. I have designed each book as the **missing manual** you wish you had to get the most value out of these components and to provide some guidance in using them for **open-ended science experiments**.

All the projects and experiments in this series are partitioned into bite-size pieces that can be completed in a Saturday afternoon. Many of them are open ended. They can each offer you many Saturdays of exploration.

I like to reserve Saturday afternoons for my electronics projects. I can get errands done in the morning and free up my afternoons to tinker around. The Arduino classes I teach at our local hacker space, Tinker Mill, are on Saturday afternoons.

We are in the middle of a revolution in access to low cost electronics, mostly available from Asia consolidators like Aliexpress or Banggood. Some of the small circuit boards you can purchase on these sites as complete, functional modules cost less

## 11.1 The name of this series of books says it all.

than if you were to buy just the main IC component through retail chains like Digikey, and for just a few dollars each.

The challenge is often how to decide what are the right components to purchase, the best way to use them and how to get the most value out of them. Some of these components, like the Arduino, the multi-function board, the LED accessory kit, the 37 sensors kit, the DMM, the scope/function generator, even the Arduino Due, Teensy and ESP32, rarely come with a *How-To Manual*.

This is one of the most important features of this series of books. I have taken the risk out of the selection process by doing my own analysis and only giving recommendations of parts I know are a good value.

Each book in this series is centered around a specific set of hardware costing just a few dollars. You will not find a collection of projects or experiments for a lower price than the ones presented in this book series. Contained in this series of books are the *missing manuals* for all of these components, dramatically increasing their value and the performance you can get out of them.

But this book series is more than about leveraging low cost electronics for projects. There is a strong focus in everything that I do and teach on learning the techniques of the professionals for conducting cool science experiments.

This is one of the really exciting consequences of the low-cost electronics revolution. Using tools you can purchase for a few dollars enables you to do experiments and perform measurements that used to cost thousands of dollars and were only available to professionals.

With a little guidance and recommendations for the parts, all provided in this book series, you too can play like the pros.

## 11.2 Welcome to the world of physical computing

More than fifty years ago, I fell in love with computers, electronics and physics. These are the basic fields used in every experiment and project in this book series. This combination of three fields has evolved to what we call today *physical computing*.

The poster child of physical computing is the Arduino microcontroller board. What defines an Arduino is that it can be programmed using the Arduino IDE. In Figure 11.1 are examples of some of the boards that are all part of the Arduino family.

 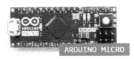 

*Figure 11.1. Some of the compatible boards in the Arduino family, courtesy of www.Arduino.cc.*

As simple as it is to write code and manipulate the microcontroller, the Arduino IDE is incredibly powerful. All the experiments and projects in this series of books can be completed using just the Arduino IDE. No other software languages or programming skills are needed.

In fact, you will learn to master the Arduino IDE by working through the projects and experiments in this book series.

The first few books in this series leverage the Arduino Uno board. This is the simplest to use, lowest cost and most robust of the Arduino boards. It is remarkable how many projects and experiments we can complete using this simple Arduino Uno board.

But when we want to do more sophisticated experiments, needing higher performance, more memory, more ADCs, DACs and I/O features, we will switch to higher performance Arduinos, the Redboard Turbo from Sparkfun, the Due, the Teensy and the ESP32. These are as powerful as any PC or microprocessor.

As with all the books in this series, I provide all the details, training and examples for anyone to get up to speed mastering these high-performance microcontrollers, using the same Arduino IDE as we learned using the simple Arduino Uno.

## 11.3 All the books in this series

***Book 1***: Arduinos without Tears: the ultimate, simplest getting started guide for Arduinos with six experiments in science and coding using nothing more than a low-cost Arduino.

***Book 2***: Science Experiments with Arduinos Using a Multifunction Board, exploring more than a dozen experiments using the Arduino and a simple add on multi-function board costing about $3. No wiring needed, just plug and play.

***Book 3***: More Science Experiments with Arduinos Using an LED Accessory Kit: exploring another dozen experiments in science and electronics using the Arduino and a simple kit of parts costing less than $3.

***Book 4***: Science Experiments with the SparkFun RedBoard Turbo, exploring more than a dozen experiments using the Sparkfun RedBoard Turbo, a high-performance data acquisition Arduino.

***Book 5***: Three Must Have Instruments for Every Lab: the missing manuals for the Arduino as a voltage source and datalogger, the Digital Multi-Meter (DMM), and the scope/function generator.

***Book 6***: Electronics You Should Have Learned in High School. Starting with the basics and establishing a firm foundation in current, voltage, resistance and power, simple circuits, voltage sources, I-V curves, diodes, schematics, measuring voltage, current, resistance, voltage dividers, capacitors and inductors.

## 11.4 About Eric Bogatin

Eric Bogatin is currently a Signal Integrity Evangelist with Teledyne LeCroy and the Dean of the Teledyne LeCroy Signal Integrity Academy, at www.beTheSignal.com . Additionally, he is an Adjunct Professor at the University of Colorado - Boulder in the ECEE dept, and technical editor of the Signal Integrity Journal.

Eric received his BS in physics from MIT in 1976 and MS and PhD in physics from the University of Arizona in Tucson in 1980.

He has held senior engineering and management positions at Bell Labs, Raychem, Sun Microsystems, Ansoft and Interconnect Devices. He has written six technical books in the field and presented classes and lectures on signal integrity world-wide.

For more than 50 years, he has followed his passions of teaching and hands-on electronics by writing, lecturing and presenting workshops on low-cost electronics to makers, DIYers, hackers and engineer and scientist wantabees around the US.

For seven years, he has presented monthly Arduino workshops at TinkerMill, our local hacker space in Longmont, CO.

Eric is a prolific author, writing engineering textbooks, project experiment books for makers and science fiction novels. In his free time, he is also an amateur astronomer and a member of the Longmont Astronomical Society and the Boulder Astronomy and Space Society.

Contact him at www.EricBogatin.com or www.HackingPhysics.com.

## 11.5 How to stay in touch

You can find more details about these and other resources on my author's page on Amazon.com and my author's website, and my Hacking Physics Journal web page.

Better yet, subscribe to the Hacking Physics Journal newsletter.

*Drop me a note sometime and let me know what you think!*

CPSIA information can be obtained
at www.ICGtesting.com
Printed in the USA
LVHW081541121020
668586LV00034B/3683